Standards in the Classroom

Standards in the Classroom

An Implementation Guide for Teachers of Science and Mathematics

Richard H. Audet
Linda K. Jordan

CORWIN PRESS, INC.
A Sage Publications Company
Thousand Oaks, California

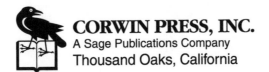

For information:

Corwin Press, Inc.
A Sage Publications Company
2455 Teller Road
Thousand Oaks, California 91320
www.corwinpress.com

Sage Publications Ltd.
6 Bonhill Street
London EC2A 4PU
United Kingdom

Sage Publications India Pvt. Ltd.
B-42 Panchsheel Enclave
Post Box 4109
New Delhi 110 017 India

Printed in the United States of America

Library of Congress Cataloging-in-Publication Data

Audet, Richard H.
Standards in the classroom: an implementation guide for teachers of science and mathematics. Richard H. Audet, Linda K. Jordan.
p. cm.
Includes bibliographical references and index.
ISBN 0-7619-3856-7 (alk. paper) — ISBN 0-7619-3857-5 (pbk.: alk. paper)
1. Science—Study and teaching—Standards—United States. 2. Mathematics—Study and teaching—Standards—United States. I. Jordan, Linda K. II. Title.

Q183.3.A1 A83 2002
507.1'073—dc21 2002034911

This book is printed on acid-free paper.

02 03 04 05 06 10 9 8 7 6 5 4 3 2 1

Acquisitions Editor: Rachel Livsey
Editorial Assistant: Phyllis Capello
Typesetter: Jeffrey Hill
Cover Designer: Michael Dubowe
Production Artist: Michelle Lee

"The oversimplification of complex issues often leads to the failure of good ideas...I see our society's tendency to oversimplify and know how fragile a great idea like standards is because people don't truly understand how to implement them or how to assess them. It makes me wonder how many great ideas have been tried but never understood fully enough to make a difference and become a lasting cog in the public education machine."

Clark, 2001
Past user of the guide

CONTENTS

PREFACE		ix
ACKNOWLEDGMENTS		xv
ABOUT THE AUTHORS		xvii
CHAPTER 1	ENGAGEMENT: THINKING ABOUT STANDARDS AND INSTRUCTION	1
CHAPTER 2	EXPLORATION: DISCOVERING NATIONAL STANDARDS RESOURCES	11
CHAPTER 3	EXPLANATION I: RESEARCHING NATIONAL STANDARDS TO CLARIFY LEARNING GOALS	27
CHAPTER 4	EXPLANATION II: ASSESSMENT IN A STANDARDS-BASED SYSTEM	47
CHAPTER 5	EXTENSION I: DEVELOPING STANDARDS-BASED CURRICULUM MATERIALS	85
CHAPTER 6	EXTENSION II: ANALYZING AND EVALUATING CURRICULUM MATERIALS	101
CHAPTER 7	EXTENSION III: CURRICULUM MAPPING	115
CHAPTER 8	EVALUATION: DISCOVERIES ABOUT STANDARDS-BASED TEACHING AND LEARNING	133
CHAPTER 9	ENDURANCE: SUSTAINING CHANGE	145
REFERENCES AND FURTHER READINGS		151
RESOURCE: PROFESSIONAL DEVELOPMENT DESIGNS		157
INDEX		176

This is dedicated to the thousands of teachers who labor daily to improve the lives of children, and to Enya, Ross, and Katie...pre-schoolers. May they discover science and mathematics within a standards-based system.

PREFACE TO THE GUIDE

Standards. The simple mention of the word incites strong emotions in all educators who are accountable for implementation of national, state, or local standards. Contrasting viewpoints abound. On one hand, the report *What Matters Most: Teaching for America's Future*, urges educators to "get serious about standards for both students and teachers" and to make teachers and teaching the linchpin of school reform (1996, p. vii). A radically different perspective is evidenced by some teachers who proclaim that, "If we just wait long enough, standards, like all other educational fads, will disappear." Are teachers wasting precious time and energy adapting their practices and curriculum to a standards-based world that may turn out to be short-lived? Or, are standards the essential ingredient for lasting reform? Based on personal experience working with teachers, the authors share the belief that when teachers derive a deeper understanding of standards, they come to appreciate their value in guiding instruction for students.

> *"Goals and standards for science education have been established; the challenge is implementing them."*
>
> **Anderson & Helms, 2001**

> *"Maybe it will go away"*
>
> **Sign in a chiropractor's office**

Standards are here to stay. Once standards were adopted, it became difficult to imagine a return to a system built on anything *but* standards. According to Reeves (1998), only two things can fill such a void: textbooks and individual teacher preferences. Neither of these represents a reasonable alternative to standards;

Standards have ramifications that permeate the entire educational system. Consequently, every teacher and administrator has a personal stake in understanding the fundamental connection between standards and the conduct of their professional lives;

Learning about standards is neither simple nor direct. Gaining a basic understanding of standards is relatively straightforward. However, translating standards into practice requires major transformations in the way that teaching and learning are viewed by educators;

Standards are most effectively learned and demystified through use. Only by continually referencing and applying new understandings about standards do these become deeply entrenched in professional practice.

Some educators believe that the standards reform movement has neither been widely embraced by classroom teachers nor produced the desired outcomes for students. This is not surprising. In our view, professional development for and by teachers is the key to achieving reform goals, and often such support has been absent at the district level. We expect that the Guide can contribute by building a support system that is required for enduring reform to occur. The Guide's goals are for its users to gain a comprehensive understanding of standards and benchmarks, become adept at using a standards-based instructional planning model, analyze curriculum through the lens afforded by the Guide's principles, and apply a process for aligning a K-12 curriculum with appropriate learning goals for students.

Three perspectives provide the foundation for the Guide. First, research has consistently shown that the final fate of any educational innovation is decided when the door to the classroom closes. As the ultimate arbiters of classroom conditions, teachers are the cornerstones of successful reform. Significant changes for students are unlikely to occur unless and until teachers' attitudes and practices change. Second, changing one's teaching style can be as difficult as changing one's learning preferences. Significant, lasting change is both daunting and challenging. The final point is that professional development provides the critical link between teacher change and successful reform. Giving teachers the opportunity to experiment with new practices in a genuine and meaningful classroom context may be the most effective method for challenging preconceived ideas about the importance of change.

PURPOSE OF THE GUIDE

Most practicing teachers prepared to teach under a different set of instructional and curricular principles than is advocated today. The advent of standards gave teachers a vision for the ends of education. The standards-based curriculum they design or select and the instruction they provide are the means for attaining these desired goals with students. In this book, we use the expression "standards-based reform" to describe the broadest educational context wherein almost everything that is associated with teaching and learning is referenced to standards. Clearly defined K-12 content standards for students and pedagogical standards for teachers affect all facets of education, from teacher preparation to the daily events that occur in classrooms.

Before beginning to write this Guide, we wondered if it was possible to create professional development tools that could successfully support educators whose goal was to implement standards-based curricula, teaching, assessment, and learning practices. Our conclusion was that such materials could be designed. But, for these tools to be effective, they had to be sensitive to the user's needs and fully embedded in a familiar classroom context. The document had to be clearly written, easy to use, realistic, and practical, and it had to employ recognizable professional development language. Teachers would only accept the Guide if the required investment in time and effort produced complementary outcomes, such as noticeable improvement in student achievement. We believe that the Guide is faithful to these concerns.

Materials in the Guide provide a suitable professional development framework for small study groups that are learning together and other formal professional development scenarios. However, a motivated classroom teacher should be able to follow the outline without assistance from an outside consultant. The amount of time spent with the Guide depends upon the needs and prior experience of the audience. We have identified several Learning Paths in the *Professional Development Designs* section that illustrate how materials from the Guide can be sorted, rearranged, and compiled to create customized formats that meet the needs of a variety of audiences.

The Guide covers the full range of issues and topics that are relevant for implementing content standards in the classroom. The focal points of the Guide are as follows:

- Understanding principles of teacher change
- Teaching and learning in a constructivist environment
- Understanding the standards-based reform movement
- Using national standards resources to better understand state and local standards
- Developing performance assessments linked to standards
- Using standards to guide instruction and the selection of curriculum materials
- Analyzing a curriculum through principles embedded in the Guide
- Applying curriculum mapping to align a curriculum with standards

To the maximum extent possible within a text format, the Guide employs a constructivist approach for discovering how to implement standards. For example, background reading is minimal. Instead, active engagement in guided discovery enables users to develop a deep personal understanding of standards-based reform issues. Incorporating such activities as surveys, graphic organizers, questionnaires, focused readings, Likert scales, rubrics, and Venn diagrams makes the Guide engaging and instructive. Follow-up reflections linked to the learning activities help users to monitor their progress toward mastery of key concepts and applications.

THE LEARNING CYCLE

The Guide uses a five-stage learning cycle model developed by the Biological Sciences Curriculum Study (1993) to create a professional development framework. The learning cycle approach to instructional delivery supports a learning environment that is carefully aligned with ideas about conceptual change. Each step in a learning cycle serves a different purpose and incorporates specially structured activities. This is why we believe that completing all five sections of the book's learning cycle produces the greatest impact. Figure P.1 on page xiv illustrates the Guide's conceptual framework.

Engagement: These activities examine personal beliefs and understandings about change and standards and their implications for reform.

Exploration: This section gives users a working familiarity with national standards resources. The exploration activities illustrate how developing a comprehensive information base can enhance standards-based teaching decisions.

Explanations: These chapters foster a deeper understanding of K-12 content standards and introduce contemporary approaches for assessing student performance. The content crosswalk and content clarification processes may present new ideas for teachers. When combined, these procedures demonstrate how conducting a research-based study of a standard can serve as a tool for implementing standards in the classroom. The assessment chapter introduces important ideas about evaluating student work from a standards perspective.

Extensions: The Guide includes three separate extensions, each aimed at achieving a different outcome. In the first, users apply their growing understanding of assessment to inform decisions about curriculum and teaching. A carefully sequenced standards-based instructional development model is introduced. Chapters 1-4 are prerequisites for successfully designing instruction according to this approach.

The second extension, Chapter 6, is tailored to users who, like most teachers, will not write curricula, but rather will make decisions about curriculum adoption for their individual classes, districts, or states. We have included a recommended Learning Path for preparing to analyze and evaluate a curriculum in terms of how carefully it adheres to standards-based reform principles.

In Chapter 7, users apply their knowledge of reform principles to systematically map existing curricula. The information derived from this analysis can be used to develop a comprehensive K-12 curriculum framework that is aligned with standards.

Evaluation: Throughout the Guide, users frequently reflect on their guided discovery experiences as a method for monitoring their progress toward success. Chapter 4 introduces the key principles of standards-based assessment. Chapter 8 includes final evaluations for the major components of the Guide.

To summarize, the first four chapters of the Guide aim at building confidence and knowledge of standards and contemporary assessment practices. Extensions in Chapters 5-7 provide opportunities to apply these new understandings for a variety of targeted purposes. Everything in the Guide ultimately targets the first extension, Developing Standards-Based Curriculum Materials. Mastering these skills is essential for success because this foundation establishes a critical and direct link between standards and their impact on student achievement.

Ongoing professional development offers the key for sustaining any changes that result from using the Guide. Chapter 9 introduces a framework for future study that combines study groups, action research, and examination of student work.

EXPECTATIONS FOR USERS OF THE GUIDE

The authors believe that the best examples of professional development are organized around collaborative inquiry. The Guide supports the collective or individual efforts of teachers who want to implement standards-based reform practices in their classrooms. The approaches work equally well with national, state, or local standards. Although the Guide principally targets science and mathematics standards, we believe that the flexibility of its approach should allow it to be adapted to other content areas, such as history/social studies and English language arts. The only additional materials required to successfully use this Guide are the national, state, and/or local standards and relevant standards resource documents.

Professional educators who successfully complete the Guide can reasonably expect to accomplish the following:

- Understand the basic principles of standards-based reform
- Know and correctly use terminology associated with standards
- Become practiced users of national standards resource documents
- Apply national standards resource documents to clarify local and state standards and curriculum frameworks
- Determine the specific intent of content standards through research
- Use a content clarification process to inform the selection of instructional materials and teaching practices
- Understand and apply the principles of curriculum congruence
- Design performance-based assessment strategies
- Recognize how information about standards support fundamental decisions about curriculum, assessment, and instruction
- Implement a standards-based instruction model
- Evaluate the features of curriculum materials in terms of standards-based reform principles
- Apply new understandings to improve student achievement
- Align a K-12 curriculum framework to standards

A willingness to allocate considerable time, a strong commitment to carefully researching standards, and openness to change are essential ingredients for successfully using the Guide. Completing this process can prepare you to effectively implement standards-based practices and may serve to transform your ideas about teaching and learning.

FIGURE P.1 CONCEPTUAL OVERVIEW OF THE GUIDE

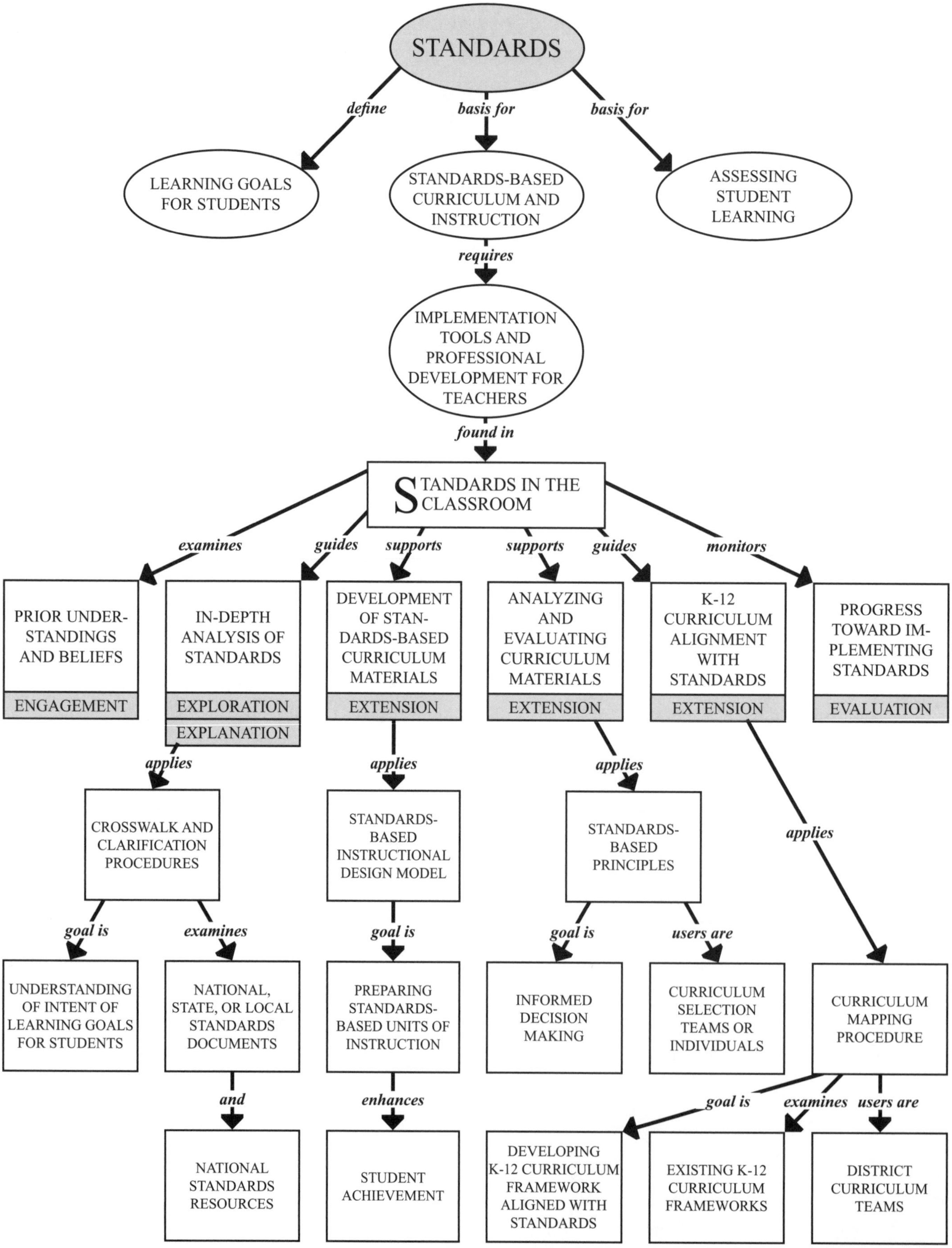

ACKNOWLEDGMENTS

Einstein defined insanity as "doing the same thing over and over again and expecting different results." This was essentially how we viewed the status of the standards-based reform agenda when we began writing this book. The hoopla over the promulgation of educational standards had subsided, but the uproar over federal and state mandated testing was reaching a crescendo.

For the history of the reform movement to have followed a smoother and more direct course, the focused professional development required for effective implementation of standards would have to have been delivered at some time between the development of standards and the onset of "standards-based" testing. In most instances, this hadn't happened. In case after case, teachers related how they were being held accountable for delivering the type of standards-based instruction for which they often lacked even rudimentary preparation.

Yet at the same time that teachers were expressing these feelings of abandonment, we were receiving urgent calls for assistance from school administrators expressing a strong desire to help teachers gain the necessary skills for implementing standards. What puzzled us was that callers were extremely vague about the kind of services they were requesting. These people seemed to realize that something had to be done, but were uncertain about potential solutions to the problem. It became evident that few effective and teacher-friendly professional development models and mechanisms existed for helping educators implement standards-based curriculum and instruction in their classrooms. We developed the learning activities in this Guide to fill this perceived gap.

Our goal was to forge a new synthesis of the major topics, concepts, approaches, and issues that revolve around standards-based reform of education. We tried to find ways to integrate disparate ideas and principles about standards into newer, clearer, and more useful configurations. We also wanted our materials to be presented and organized into a highly graphic and logical format that teachers would appreciate.

To the extent that the Guide fulfills these goals, we are deeply indebted to the many people who influenced the course of our professional lives and thus, indirectly, the composition of this book. We wish to acknowledge those individuals who we consider as joint authors. First and foremost, we thank Jeffrey Hill for his unwavering commitment to this project. His creative talents and ability to translate our ideas into graphical representations give the Guide its unique appearance. Those who know the work of Mary Ann Brearton at Project 2061 will recognize her impact on this book. Much of our own professional growth occurred under her wise tutelage. Another strong mentoring influence stemmed from our affiliation with the WestEd Science Education Leadership Academy. We like to imagine that the late Susan Loucks-Horsley and her associates would share our joy and excitement for this project.

Teachers and administrators who participated in workshops (using less-than-developed materials!) provided us with valuable feedback that we incorporated into the Guide. We appreciate the patience and contributions of educators from the Warwick (RI), Bristol/Warren (RI), and Bridgewater/Raynham (MA) Public Schools; the Appalachia Education Laboratory's Foundations for Gateway Success Team Leaders and Trainers; the HELMSS NSF middle school project; the Rhode Island Department of Education's Standards Institute; the Tennessee Textbook Review Committee; and the Tennessee K-12 Standards-Based Unit Sampler Development group. We are grateful to our manuscript reviewers, Lisa Audet, Pat Bowers, Jack Rhoton, Nancy Love, Linda Jzyk, Bruce Marlowe, Susan Mundry, Russ Rapose, Kathy Steeves, and especially, Barbara Walton-Faria. Your contributions were significant, your encouragement sustaining. Special thanks go to Barbara Zahm, who suggested the need for Chapter 6. Without her advice, we might not have added this valuable extension to our original plan.

Rachel Livsey, Phyllis Cappello, and Scott Hooper made up the excellent project team organized by Corwin Press. Their approach to working with authors was a perfect match to our need for ongoing encouragement. Finally, we want to recognize the special influence drawn from the loving support of our family members.

The contributions of the following reviewers are also gratefully acknowledged:

Cathy Hewson
Mathematics Teacher
Rio Mesa High School
Oxnard, CA

Joseph H. Peake, Jr.
Director
Central Coast School Leadership Center
Santa Barbara, CA

Marti Richardson
President-Elect
Board of Trustees
National Staff Development Council
Knoxville, TN

Theresa R. Rouse
Assessment and Program Evaluation Coordinator
Monterey County Office of Education
Salinas, CA

Cindy Harrison
Director, Staff Development
Adams 12 Schools
Thornton, CO

Arlene Delloro, Ed.D.
Principal
Montebello Elementary School
Suffern, NY

ABOUT THE AUTHORS

Richard H. Audet is an associate professor of Science Education at Roger Williams University in Bristol, Rhode Island. His primary area of interest is teacher preparation. He consults nationally on the topic of standards-based education in science. He holds undergraduate and graduate degrees in biology from Providence College and a doctorate in science education from Boston University. He taught high school biology for more than 20 years.

Linda K. Jordan is the K-12 Science Consultant for the Tennessee State Department of Education, Division of Curriculum and Instruction in Nashville. She provides technical assistance to local education agencies throughout her state and coordinated the recent revision of Tennessee's state science standards. She is a board member of the Council of State Science Supervisors and has extensive national and statewide experience as a professional development provider. She is a former high school biology teacher. Her bachelor's, Master of Science, and Ed.S. degrees were earned at the University of Tennessee, Knoxville.

ENGAGEMENT: Thinking About Standards for Instruction

CHAPTER 1

In the learning cycle approach for organizing instruction, the way that a person is initially engaged in the learning experience is critically important. From a constructivist perspective, every learner enters the classroom processing a vast store of information. Activating an individual's prior understanding without regard to the veracity of this preliminary knowledge is viewed as the essential precursor to learning. This process creates a context within which the learner may then master a concept at the required depth. Engaging a student with an idea is analogous to stirring a person's intellectual pot. The experience sets off an initial response that promotes a learning environment that is both motivational and connected with what the learner already knows. Chapter 1 opens with an exploration of prior beliefs about change in anticipation of the learning experiences that will follow in the Guide.

> *"The education field is subject to many fads, and what counts as a good idea varies over time and across locations. At present, most people are persuaded that the key to educational improvement lies in developing a coherent and integrated system for governing education...based on the same set of ideas. These ideas have come to be called standards."*
>
> **Kennedy, 1998**

	LEARNING GOALS	ACTIVITIES
ENGAGEMENT	Develop motivated users of the guide.	1.1: My Beliefs and Attitudes About Change
	Activate prior beliefs about change.	
	Examine personal beliefs and attitudes about standards.	1.2: Beliefs and Attitudes About Constructivism and Standards-Based Learning
	Explore ideas about constructivist thinking.	

SUMMARY REFLECTION	*Thinking About an Issue From a Right-Angle Perspective*

ENGAGEMENT 1.1 My Beliefs and Attitudes About Change

DIRECTIONS

The sole purpose of this belief inventory is to prompt your thinking about the broad process of change. The actual score you give yourself is really not important. Read each item carefully. Quickly record your response by placing a mark along the five-point scale described below that best expresses your initial reaction to the statement. There is a reflection for you to complete after you finish this survey.

> *"On the one hand, we have the constant and ever expanding presence of educational innovation and reform…on the other hand, we have an educational system that is fundamentally conservative…more likely to retain the status quo than to change."*
>
> **Fullan, 1993**

0	This statement is irrelevant to my ideas about change.
1	This statement contradicts my ideas about change.
2	This statement is somewhat consistent with my ideas about change.
3	This statement is very consistent with my ideas about change.
4	This statement is highly consistent with my ideas about change.

Change is a highly personal thing.	⓪ ① ② ③ ④	When you change, you lose something of yourself.	⓪ ① ② ③ ④
Change and stress go hand in hand.	⓪ ① ② ③ ④	I need to know that where we are headed is better than where we've been.	⓪ ① ② ③ ④
I need to understand the implications of change before I can seriously think about changing the way I do things.	⓪ ① ② ③ ④	Change is good.	⓪ ① ② ③ ④
Change is chaotic.	⓪ ① ② ③ ④	I tense up when people around me talk about change.	⓪ ① ② ③ ④
I prefer change that is self-initiated.	⓪ ① ② ③ ④	Change is best when it happens incrementally.	⓪ ① ② ③ ④
People need nurturing and support if they are to change.	⓪ ① ② ③ ④	Change is a necessary part of personal growth and development.	⓪ ① ② ③ ④
I am too busy to change.	⓪ ① ② ③ ④	Some people like to change simply for the sake of changing.	⓪ ① ② ③ ④
I consider myself to be a risk-taking, innovative person.	⓪ ① ② ③ ④	People need sufficient time to change.	⓪ ① ② ③ ④
Change…who needs it?	⓪ ① ② ③ ④	People change more easily than institutions.	⓪ ① ② ③ ④
People differ widely in their willingness and ability to change.	⓪ ① ② ③ ④	Change is a process, not an event.	⓪ ① ② ③ ④

REFLECTION 1.1 My Beliefs and Attitudes About Change

1. Analyze how you rated yourself on this belief inventory. What do the overall results suggest about your level of comfort with change? Does the survey accurately portray your general beliefs and attitudes about change?

> *"Education reform that makes a difference for students requires teachers and principals to respond in new ways to the need for change and to rebuild the very foundation of their thinking about teaching and learning."*
>
> **National Staff Development Council, 2000**

2. Assume that a major change initiative is about to begin at your school. What does your analysis of the responses on the belief inventory suggest about how you might react?

3. What is your reaction to the following statement made by Michael Fullan (1993)? He wrote that, "Almost every important learning experience that we have ever had has been stressful…This means that the ability to suspend beliefs, take risks, and experience the unknown [is] essential to learning" (p. 16).

REFLECTION 1.1 My Beliefs and Attitudes About Change

4. In discussing his model of teacher change pictured below, Thomas Guskey (1986) states, "Significant change in teachers' beliefs and attitudes is likely to take place only after changes in student learning outcomes are evidenced" (p. 7). What is your reaction to this statement?

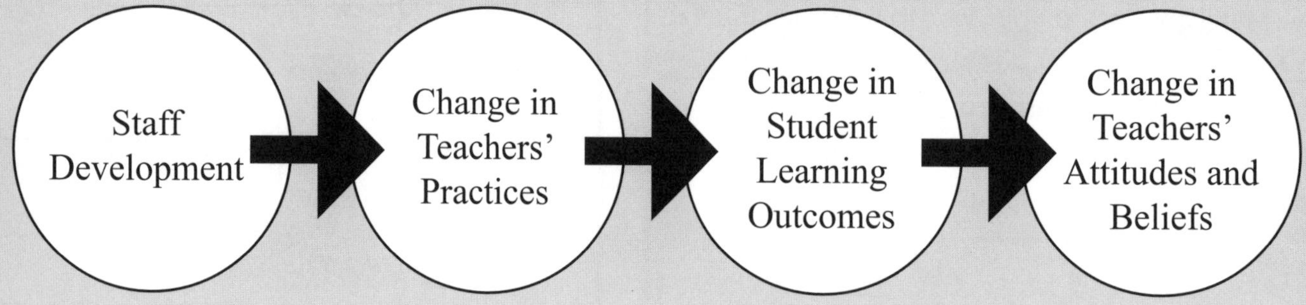

ENGAGEMENT 1.2 Beliefs and Attitudes About Constructivism and Standards-Based Learning

In Engagement 1.2, you will complete an exercise that combines self-assessment with personal goal setting. Where do you presently fall on the scale below that contrasts the characteristics of constructivist, standards-based learning environments with more traditional classrooms?

> *"Fundamental reform...requires significant changes in teachers' values and beliefs about (science) education practice."*
>
> **Anderson & Helms, 2001**

DIRECTIONS

Think of these statements as representing two contrasting extremes along a continuum. Place an O to mark the spot on this continuum where you presently find yourself and an X to target the place where you hope to be after working through this Guide. Sharing the results of your personal inventory is optional.

EXAMPLE:

Emphasis on manipulative curriculum materials	...........X..............................O..	Emphasis on text materials

1	Varied instructional approaches	....	Lecture as principal method of instruction
2	Teacher as mediator of learning environment	....	Teacher as disseminator of information
3	Teacher working with groups of students	....	Teacher generally at the front of the class
4	Collaboration among teachers	....	Teachers working in isolation
5	Support through ongoing professional development	....	Professional development a personal initiative
6	High standards for all students	....	High standards for the best and brightest
7	High expectations for all students	....	Different expectations for different students
8	Not all standards reached by all students at same rate	....	Expected progress identical for all students
9	Prior knowledge of student critical to learning	....	Students' minds are blank slates
10	Students as active learners	....	Students as recipients of information

ENGAGEMENT 1.2 Beliefs and Attitudes About Constructivism and Standards-Based Learning

11	Collaboration among students	....	Students working alone
12	Students assume greater responsibility of work	....	Teacher is the worker
13	Student questions highly valued	....	Questions provided by teacher
14	Curriculum decisions based on standards	....	Curriculum driven by coverage of text material
15	Emphasis on student learning	....	Emphasis on teaching
16	Focus on understanding	....	Focus on acquiring factual knowledge
17	Clear expectations for students	....	Expectations for students not openly stated
18	In-depth study of fewer topics	....	Coverage is broad and lacks depth
19	Integration of content areas	....	Content areas separate
20	Experimental, inquiry-based learning	....	Teaching driven by texts and workbooks
21	Curriculum adapted for individual students	....	One curriculum for all
22	Multiple sources of evidence for student learning	....	Few sources of evidence for student learning
23	Continuous monitoring of student progress	....	End-of-unit tests
24	Numerous opportunities for self-assessment	....	Little self-assessment
25	Student progress reported in terms of standards	....	Traditional grade reporting
26	Heterogeneously grouped classes	....	Ability tracking of students
27	Student-centered instruction	....	Teacher-centered instruction
28	Students guided toward inquiry	....	Lock-step approaches

REFLECTION 1.2 Beliefs and Attitudes About Constructivism and Standards-Based Learning

1. How do you relate the quote by Dewey to your personal teaching philosophy?

"The teacher is a guide and director; he steers the boat, but the energy that propels it must come from those who are learning. The more a teacher is aware of the past experiences of students, of their hopes, desires, chief interests, the better will he understand the forces at work that need to be directed and utilized for the formation of reflective habits."

Dewey, 1910

2. Does moving toward a standards-based curriculum require a new style of teaching? If yes, then what does this instruction look like? If no, then why not?

3. Lynch (1997), in a paper from the *Journal of Research in Science Teaching*, said that "the alignment of teachers' core beliefs with a curriculum's conceptual framework (standards) is crucial to full implementation" (p. 3). Is it necessary for a teacher to be fully committed to standards in order to successfully implement them? How do you think Guskey (see p. 4) would react to this statement?

REFLECTION 1.2 Beliefs and Attitudes About Constructivism and Standards-Based Learning

4. What do you currently see as the major challenges and obstacles to implementing standards-based instruction?

5. How might your beliefs and attitudes about change that you identified in Engagement 1.1 influence your efforts to move toward instruction that is more closely aligned with standards?

SUMMARY REFLECTION

CHAPTER 1

Thinking About an Issue From the Right-Angle Perspective

DIRECTIONS

In Box 1, list all the feelings that you have about standards and benchmarks. In Box 2, list what you know. Write a summary statement about standards and benchmarks in Box 3.

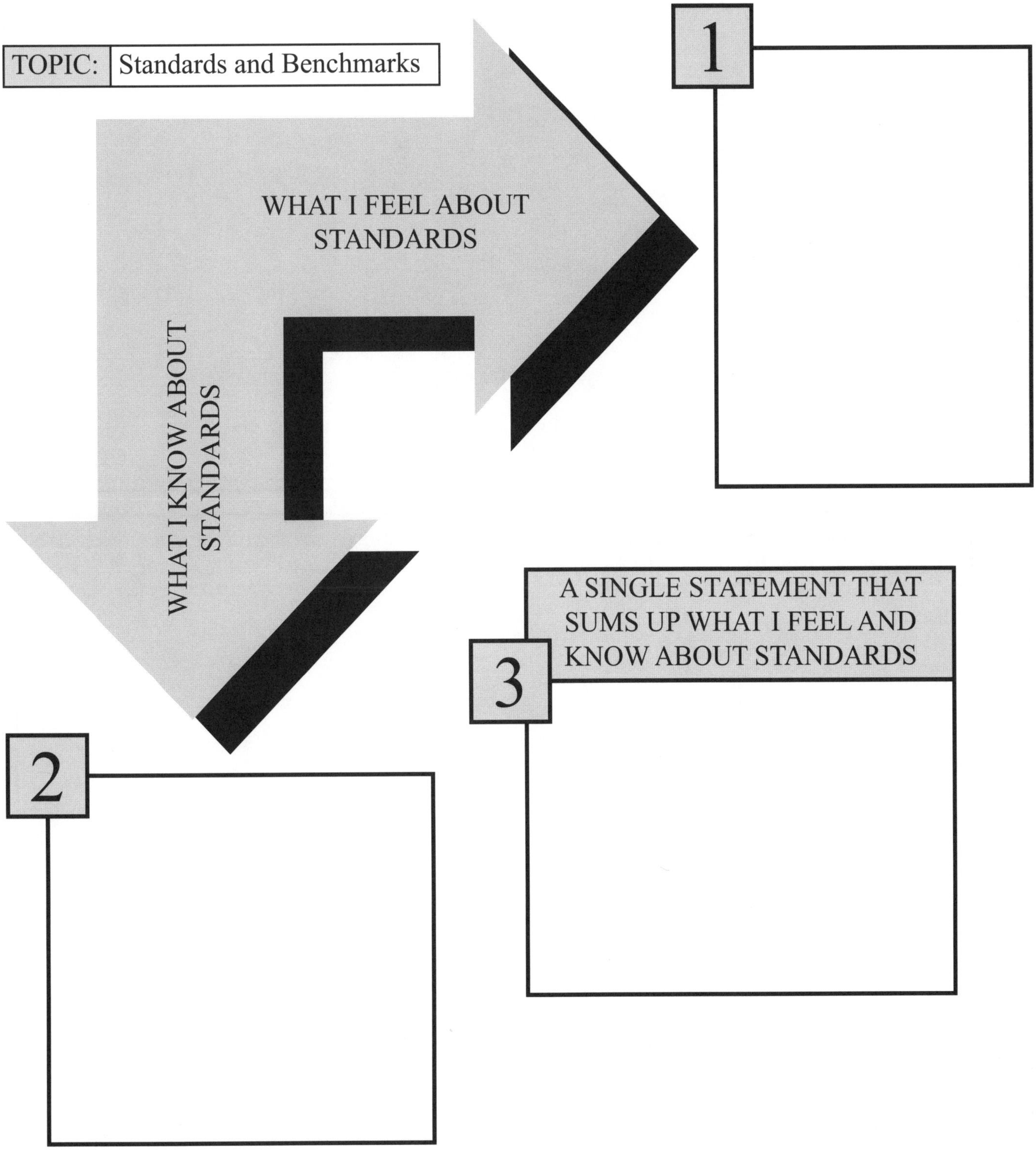

SUMMARY REFLECTION

CHAPTER 1

Thinking About an Issue From the Right-Angle Perspective

4. What is the essential difference between a person's beliefs about a topic and his or her knowledge and understanding of this topic?

5. Why is it important for people to be able to clearly distinguish between their beliefs and knowledge about a topic?

EXPLORATION: Discovering the National Standards Resources

CHAPTER 2

Explorations are student-centered experiences that are not generally intended to enhance understanding or correct misunderstandings. The primary purpose of activities in this stage of the learning cycle is to trigger self-awareness of prior knowledge. This establishes cognitive conditions for the insights that will hopefully follow. The explorations in the Guide will enable you to gain a deeper appreciation of the quality and potential importance of national standards documents.

> *"Every adult will be literate and possess the knowledge and skills necessary to compete in a global economy and exercise the rights and responsibilities of good citizenship."*
>
> **U.S. Department of Education, 1991**

	LEARNING GOALS	ACTIVITIES
EXPLORATION	Assess prior understanding about standards.	2.1: Developing a Common Language About Standards
	Develop a language for standards.	2.2: Internet Discovery Guide
	Become familiar with the national standards resources.	2.3: Getting to Know the National Standards Resources

SUMMARY READING AND REFLECTION	*Realizing the Promise of Standards-Based Education* (Schmoker & Marzano, 1999)

EXPLORATION 2.1 Developing a Common Language About Standards

Because this is an exploration, the goal is not to provide correct explanations but to activate what you already know about terminology associated with the standards reform movement. Remember, today's understanding is merely a prelude to what you will become better acquainted with tomorrow.

> *"The dirty little secret is that we already have informal national standards because so much of our teaching is based on similar lessons, taught in a similar way using roughly the same textbooks and multiple-choice tests."*
>
> **Zuckerman, 1996**

DIRECTIONS

1. Carefully review each of the terms in the left-hand column of Exploration 2.1, on the following page.

2. In the Rating column, score each term on a scale of 1 to 3 according to your present level of familiarity.

1	No understanding of the term
2	Limited understanding of the term
3	Thorough understanding of the term

3. For each term that receives a score of 2 or 3, you should search the right-hand column until you find the corresponding *descriptor*. Ignore those terms that receive a score of 1.

4. When you find the descriptor that matches the term, place the corresponding letter next to this number. Do not deliberate too long on any one item.

5. As with most educators today, you might be unable to make some of these connections. This should not alarm you. After completing all five major pieces of the Guide, the expectation is that you will have developed a strong personal understanding of the terms commonly used in professional conversations about standards.

6. An answer key is provided at the end of the Reflection for this section.

EXPLORATION 2.1 Developing a Common Language About Standards

TERM	RATING (1-3)				DESCRIPTOR
Benchmarks		1		A	A particular subject or discipline.
Best practice for teaching and learning		2		B	Creation of learning environments that make necessary learning accommodations for all learners.
Content area		3		C	Broad ends toward which the curriculum is directed. Foundation for the development of standards and benchmarks.
Content standard		4		D	Components of the standards. Specify the learning goals for a particular grade level, grade cluster, or course.
Curriculum integration		5		E	Scoring guide that describes the criteria on which a piece of student work will be evaluated.
Goals		6		F	Research-based approaches to effective teaching and learning.
Inclusion		7		G	Use of connections among different content areas to promote student learning.
Inquiry		8		H	Test instrument used to determine how students perform on a particular set of standards.
Learning goal or expectation		9		I	Detailed statement about what students should know and be able to do. A standard can include several of these.
Multiple measures		10		J	Approach to change in which every component of the educational system is addressed.
Performance assessment		11		K	Approach to teaching and learning build around the exploration of student ideas.
Performance indicator		12		L	General statement that identifies the knowledge, skills, and dispositions developed through instruction in a specific content area.
Performance task		13		M	An established level of achievement, quality of performance, or degree of proficiency.
Prior knowledge		14		N	Activity, exercise, or problem that enables students to demonstrate what they know or are able to do.
Rubric		15		O	Assessment approach that uses many data sources to assess student performance.
Self-assessment		16		P	Assessment that requires students to construct a response, create a product, or perform a demonstration.
Standards-based assessment		17		Q	Any technique that is used to monitor one's own progress.
Systemic reform		18		R	What the learner already knows about a topic or standard.

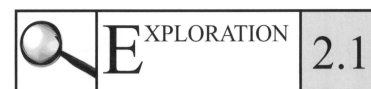

REFLECTION 2.1 Developing a Common Language About Standards

1. What are your initial thoughts after having completed Exploration 2.1?

"Standards, quite simply, refer to a level of quality. They provide a goal towards which one aspires…in this case, the goal is scientific literacy."

National Science Teachers Association, 1998

2. On a scale of 1 to 10, how would you rate your general familiarity with terms that are commonly used to discuss standards?

3. People who use this Guide are apt to have had different levels of understanding for these terms? What are some possible explanations for these variations?

REFLECTION 2.1 Developing a Common Language About Standards

4. Do you think that it is necessary to be familiar with all of these terms in order to successfully implement standards in your classroom? Explain your answer.

5. In an Exploration, why is it relatively unimportant that you produce the correct answers for a task like the one you just completed?

6. As you completed this task, did any of your ratings change? How can you explain this?

ANSWER KEY

You should consult these answers only after you have had the opportunity to complete and reflect upon Exploration 2.1.

1(D); 2(F); 3(A); 4(L); 5(G); 6(C); 7(B); 8(K); 9(I); 10(O); 11(P); 12(M); 13(N); 14(R); 15(E); 16(Q); 17(H); 18(J).

EXPLORATION 2.2 Internet Discovery Guide

Point your favorite Web browser to:

www.statestandards.com OR
www.edstandards.org

> *"One of the current problems with the standards movement is that there are several categories of standards to consider."*
>
> **Foriska, 1998**

The Internet is an invaluable resource for anyone who is interested in learning about standards and standards-based instruction. The sites listed above provide a wealth of information about national, state, and even local standards and contain links to standards-based instructional materials. If the above Web sites are inactive, then you should go directly to the state department of education Web site and national standards Web sites to locate the standards documents that are relevant to your teaching.

As this 50-state perspective illustrates, the standards reform movement is a nationwide phenomenon. In the United States, there are no mandated national curricula. Therefore, the national standards you will examine in Exploration 2.3 do not represent a curriculum. Rather, these documents offer guidance for educators to develop corresponding curricular material and present ideas about instruction designed to meet the needs of *all* students.

1. Who sponsors these Web sites?

2. Are the standards for your state directly accessible through the above Web sites? If not, how can you access these documents on the Web?

EXPLORATION 2.2 Internet Discovery Guide

3. Take some time to review the *state* standards for the content area and/or grade level that you teach. How would you describe your current level of familiarity with these materials?

4. What additional resources are available through these Web sites?

5. What are some potential uses for the resources that are accessible through these Web sites?

6. What are the Web sites where you can access the *national* standards for your content area? What additional resources are available at these Web sites?

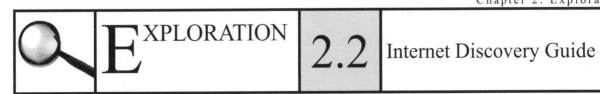

EXPLORATION 2.3 Getting to Know the National Standards Resources

During the 1990s, professional organizations such as the American Association for the Advancement of Science, the National Council of Teachers of Mathematics, the National Council of Teachers of English, and the National Council of Social Studies Teachers developed national standards for each of their respective disciplines. These federally funded projects involved thousands of people from across the educational spectrum.

> *"Standards provide a vision of what science teachers need to understand and what they need to do to assure adequate learning experiences for all students."*
>
> **Bybee, 1997**

Soon after the publication of national standards, individual states joined the reform movement and developed their own content standards and benchmarks. Although state documents are rich sources of information about standards-based reform for teachers, they seldom offer the depth of information and research perspective found in the national resources.

Groups that were originally involved in developing national standards continue to publish materials that clarify and elaborate upon the original standards documents. These publications provide valuable resources for teachers who are committed to implementing standards-based curriculum and instructional practices.

In Exploration 2.3, you carefully examine documents published by major national professional organizations to develop a clearer understanding of the philosophical and theoretical foundations of the standards reform movement. Discovery Guides for science and mathematics are provided here for you to complete an exploration of each of the national documents. Later, you will use these same resources to gain a better understanding of the state and/or local standards that are most closely linked with your own professional practice.

EXPLORATION 2.3

Science Discovery Guide: *Science for All Americans*

These are sources of information that you need to complete the Discovery Guide. You should complete each of the discovery tasks and identify the page(s) where you located the information.

- American Association for the Advancement of Science. (1990). *Science for all Americans*. New York: Oxford University Press.
- www.project2061.org

PAGE(S)	1. What national organization spearheaded the preparation of this document? Who participated in writing it? When was it published?
PAGE(S)	2. Why do you think that the particular title, *Science for All Americans*, was chosen for this book?
PAGE(S)	3. When *SFAA* uses the expression "science literacy," what is the intended meaning?
PAGE(S)	4. Each chapter of *SFAA* is built around a set of Recommendations. Select a single idea from a chapter that relates directly to what you teach. Summarize the major points in the section. Do you agree with what *SFAA* recommends all high school graduates should know about this concept? Explain.
PAGE(S)	5. Find the chapter titled Common Themes. Describe the four big ideas that pervade all of science, mathematics, and technology.
PAGE(S)	6. Which of the Principles of Learning do you regard as the most important? Why?
PAGE(S)	7. What are the major differences between the type of teaching and learning suggested in this document and what you typically see and use in your school?
PAGE(S)	8. What chapters in *SFAA* would be of special interest to a person with an interest in the history of science, mathematics, and technology?

EXPLORATION 2.3 Science Discovery Guide: *Benchmarks for Science Literacy*

These are sources of information that you need to complete the Discovery Guide. You should complete each of the discovery tasks and identify the page(s) where you located the information.

- American Association for the Advancement of Science. (1993). *Benchmarks for science literacy*. New York: Oxford University Press.
- www.project2061.org

PAGE(S)	1. What national organization spearheaded the preparation of this document? Who participated in writing it? When was it published?
PAGE(S)	2. Why do you think that the particular title, *Benchmarks for Science Literacy*, was chosen for this book?
PAGE(S)	3. Select a chapter between three and six. What are the principal differences between the content of this chapter in *SFAA* and *Benchmarks*?
PAGE(S)	4. What are the grade ranges or clusters for which benchmarks were developed? What was the rationale for benchmarking these particular grade levels?
PAGE(S)	5. Some people use the expressions "benchmarks" and "learning goals" interchangeably. Can you explain why this happens?
PAGE(S)	6. Chapter 15 is titled The Research Base. What was the connection between educational research and the development of *Benchmarks*?
PAGE(S)	7. Refer to Chapter 15, The Research Base. Select a section that references the chapter that you chose for Item 3 above. What are the principal research findings about teaching and learning that are associated with this major idea?

EXPLORATION 2.3

Science Discovery Guide: *National Science Education Standards*

These are sources of information that you need to complete the Discovery Guide. You should complete each of the discovery tasks and identify the page(s) where you located the information.

- National Research Council. (1996). *National science education standards*. Washington, DC: National Academy Press.
- http://www.nap.edu/books/0309053269/html/

PAGE(S)	1. What national organization spearheaded the preparation of this document? Who participated in writing it? When was it published?
PAGE(S)	2. What are some of the essential differences between the *NSES* and *Benchmarks*? The major similarities?
PAGE(S)	3. What are the eight content areas for which science standards were developed?
PAGE(S)	4. What are the grade ranges or clusters for which the national science standards were developed? What was the rationale for grouping these particular grade ranges?
PAGE(S)	5. The *NSES* treats Inquiry as a content standard. What do you think about designating inquiry as a separate science content area?
PAGE(S)	6. Are the *NSES* and the *Benchmarks* the same as a science curriculum? Explain.
PAGE(S)	7. *NSES* includes standards for science teaching while *Benchmarks* does not. What are the major differences between the type of teaching and learning suggested in the *NSES* document and what you typically see and use in your school?
PAGE(S)	8. What does it mean when standards documents refer to the idea of "systemic reform"?

EXPLORATION 2.3

Mathematics Discovery Guide: *Principles and Standards for School Mathematics*

These are sources of information that you need to complete the Discovery Guide. You should complete each of the discovery tasks and identify the page(s) where you located the information.

- National Council of Teachers of Mathematics. (2000). *Principles and standards for school mathematics.* Reston, VA: Author.
- www.nctm.org

PAGE(S)	1. What national organization spearheaded the preparation of this document? Who participated in writing it? When was it published?
PAGE(S)	2. What are the grade "bands" for which mathematics standards are developed? What is the rationale for selecting these particular grade ranges?
PAGE(S)	3. Is *Principles and Standards* the same as a mathematics curriculum? Explain.
PAGE(S)	4. What are the elements of effective mathematics teaching identified in *Principles and Standards*?
PAGE(S)	5. What are the major differences between the type of teaching and learning suggested in this document and what you typically see and use in your school?
PAGE(S)	6. What does *Principles and Standards* say about the need for learning mathematics with understanding?
PAGE(S)	7. What is the equity principle advocated in *Principles and Standards*?
PAGE(S)	8. How does *Principles and Standards* address the issue of assessment?
PAGE(S)	9. Examine Chapter 3. What important understanding about student learning can be gained from a careful reading of this material?

SUMMARY READING AND REFLECTION CHAPTER 2

Realizing the Promise of Standards-Based Reform

This is a widely read article about the impact of standards on educational reform. You should complete the Sentence Stems while you read this article. This paper can be downloaded from the Web. If the paper is not available in electronic form at this Web site, the reference is: Schmoker, S. & Marzano, R. J. (1999). Realizing the promise of standards-based education. *Association for Supervision and Curriculum Development*,

www.ascd.org/readingroom/edlead/9903/extschmoker.html

Authors:	
Source:	
General Topic:	

From My Perspective...

1. The authors' main points were...

2. The most memorable quote was...

3. The most surprising part of the article was...

4. My major discovery was...

5. What's confusing is...

6. I'd like to know more about...

EXPLANATION I: Researching National Resources to Clarify Learning Goals

CHAPTER 3

In a learning cycle, the explanation stage occurs when new information is presented. Explanation activities focus on content knowledge that adds to, coincides with, or challenges a learner's present understanding. In Chapter 3, you will investigate the *intent* of a national standard and study the learning goals that are subsumed within the standard. You will also use a correlation tool called a Content Crosswalk to integrate information from a variety of sources, many of which draw from the research findings on teaching and learning.

> *"Some ask if a certain curriculum meets the standards. This reveals a lack of understanding about the role of standards as outcomes. 'Meeting the standards' is correctly interpreted as students achieving the knowledge, abilities, and understandings defined in the standards." (Emphases added)*
>
> **Bybee, 1997**

Standards identify broad ends for student learning. Because they are written as general expressions, descriptions of standards offer little explicit direction for teachers in designing assessments or selecting corresponding instructional materials. In contrast, learning goals provide a much higher level of specificity for making curriculum and assessment decisions that are consistent with standards. For the purposes of the Guide, we will refer to other terms, such as benchmarks and expectations, as *learning goals*.

	LEARNING GOALS	ACTIVITIES
EXPLANATION I	Examine how a Content Clarification can foster an in-depth understanding of a standard or learning goal.	3.1: Content Clarification of a National Standard
	Construct a Content Crosswalk.	3.2: Creating a Content Crosswalk
	Recognize how national standards resources can provide information for interpreting state and local content standards.	3.3: Applying a Content Crosswalk to Clarify a Learning Goal

SUMMARY REFLECTION	*Content Clarification of Standards and Learning Goals*

EXPLANATION 3.1 Content Clarification of a National Standard

In this exploration, you seek to gain a deeper awareness of standards. You apply a content clarification procedure for examining a single national standard through the lenses afforded by the national reform documents. Such a process is important because it reveals why the developers of standards believed that this particular item or topic was important for all students to know or be able to do.

> *"A major reason for the difficulties of implementation...of education reform is that many educators simply do not understand its principles and implications, rather than not buying into the goals of reform."*
>
> **Lynch, 1997**

DIRECTIONS

1. Scaffolding is a constructivist teaching and learning strategy that uses a series of interconnected questions wherein the answer to one query creates a bridge to the next. The following set of scaffolded questions is aimed at helping you to select a particular national standard in the content area of your interest.

2. Enter the standard you selected in the Discovery Guide: Content Clarification of a National Standard (Figure 3.1 on p. 31).

3. Follow the procedure given in the next section for completing the Content Clarification of a National Standard. Of special importance to the clarification process are references to content knowledge, student learning, and aspects of instruction.

E. What is the national content standard for this topic that you would like to investigate?	
D. What are the titles of the national standards documents that include this content topic?	
C. What particular topic within this content area do you want to consider?	
B. What specific content area do you want to target?	
A. What course or grade level do you teach?	

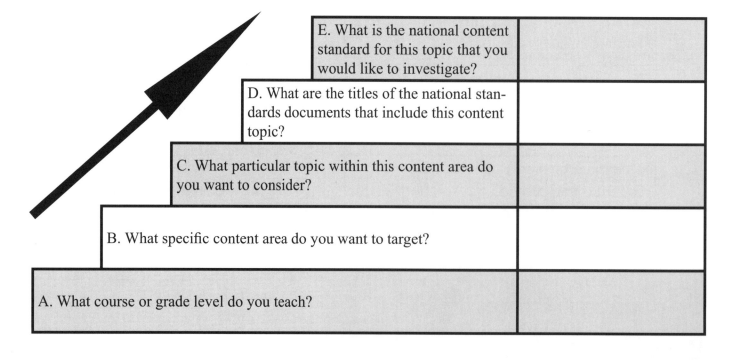

EXPLANATION 3.1 SCIENCE: Procedure for Completing a Content Clarification of a National Standard

RELATED CONTENT KNOWLEDGE

• Read sections of the chapters in *Science for All Americans* that reference this same topic and standard.

• Read the essays associated with this standard in *Benchmarks for Science Literacy*. Essays are found in the chapter introductions and as lead-ins for each set of grade cluster benchmarks. Most chapter titles and numbers in *Science for All Americans* and *Benchmarks* correspond.

• Read the sections of *National Science Education Standards* that reference this same standard or ones that it most closely approximates.

• Summarize your findings in the far left column of the Discovery Guide (Figure 3.1).

DISCOVERIES ABOUT STUDENT LEARNING

• Read the essays associated with this standard in *Benchmarks for Science Literacy*.

• Read all associated K-12 *learning goals* in *Benchmarks for Science Literacy*. These are organized according to grade clusters K-2, 3-5, 6-8, and 9-12.

• Read Chapter 15 in *Benchmarks for Science Literacy* for the educational research that references this standard.

• Summarize your findings in the center column of the Discovery Guide (Figure 3.1).

IDEAS ABOUT TEACHING TO THIS STANDARD

• Read Chapter 15 in *Benchmarks for Science Literacy* for the educational research that references this standard.

• Read the sections of *National Science Education Standards* that reference this same standard or the ones that it most closely approximates.

• Summarize your findings in the far right column of the Discovery Guide (Figure 3.1).

OPTIONAL

• Study the Strand Map in *Atlas of Science Literacy* that includes this standard.

• Figure 3.2 gives an example of a completed Content Clarification of a National Standard. *We recommend that you study this.*

EXPLANATION 3.1

MATHEMATICS: Procedure for Completing a Content Clarification of a National Standard

RELATED CONTENT KNOWLEDGE

• Read the corresponding sections in *Principles and Standards for School Mathematics* (Chapters 3-7).

• Read sections of the chapters in *Science for All Americans* that reference this same topic and standard.

• Read the essays associated with this standard in *Benchmarks for Science Literacy*. Essays are found in the chapter introductions and as lead-ins for each set of grade cluster benchmarks. Most chapter titles and numbers in *Science for All Americans* and *Benchmarks* correspond.

• Summarize your findings in the far left column of the Discovery Guide (Figure 3.1).

DISCOVERIES ABOUT STUDENT LEARNING

• Read the essays associated with this standard in *Benchmarks for Science Literacy*.

• Read all associated K-12 *learning goals* in *Benchmarks for Science Literacy*. These are organized according to grade clusters K-2, 3-5, 6-8, and 9-12.

• Read Chapter 15 in *Benchmarks for Science Literacy* for the educational research that references this standard.

• Summarize your findings in the center column of the Discovery Guide (Figure 3.1).

IDEAS ABOUT TEACHING TO THIS STANDARD

• Read Chapter 15 in *Benchmarks for Science Literacy* for the educational research that references this standard.

• Summarize your findings in the far right column of the Discovery Guide (Figure 3.1).

OPTIONAL

• Study the Strand Map in *Atlas of Science Literacy* that includes this standard.

• Figure 3.2 gives an example of a completed Content Clarification of a National Standard. *We recommend that you study this.*

NOTE: The Content Clarification of a National Standard was adapted from Project 2061. *Resources for Science Literacy: Professional Development*. 1997. Oxford University Press, and *A Leader's Guide to Bridging Maine's Learning Results With National Science Standards*. 2001. Maine Mathematics and Science Alliance. Page Keeley, Director.

FIGURE 3.1 DISCOVERY GUIDE: CONTENT CLARIFICATION OF A NATIONAL STANDARD

National Content Standard:

RELATED CONTENT KNOWLEDGE	DISCOVERIES ABOUT STUDENT LEARNING	IDEAS ABOUT TEACHING TO THIS STANDARD

FIGURE 3.2 DISCOVERY GUIDE: CONTENT CLARIFICATION OF A NATIONAL STANDARD

National Content Standard	***NSES*: Life Science Standard C - Diversity and Adaptations of Organisms** ***Benchmarks*: The Living Environment - Standards 5A Diversity of Life, 5B Heredity**	
RELATED CONTENT KNOWLEDGE	DISCOVERIES ABOUT STUDENT LEARNING	IDEAS ABOUT TEACHING TO THIS STANDARD
Elementary School Level *(NSES p. 129; Benchmarks pgs. 102, 103; SFAA pgs. 60, 61)* ➢ Millions of plant and animal species exist. ➢ Living things can be sorted into groups. ➢ Organisms have basic needs. ➢ Living things have features that help them to survive. ➢ Some characteristics are inherited; others result from interactions with the environment.	**Elementary School Level** *(NSES p. 128; Benchmarks pgs. 102, 103, 340)* ➢ Movement is used to define life. ➢ Children use mutually exclusive rather than hierarchical categories. ➢ Ideas about inheritance may contain misconceptions. ➢ Concept that organisms depend on their environment is not well developed. ➢ Students generally fail to recognize patterns among things and events.	**Elementary School Level** *(NSES p. 128; Benchmarks pgs. 102, 103)* ➢ Make direct connections with the child's world. ➢ Have students invent different ways to classify objects. ➢ Emphasize direct observations of living things and their environment. ➢ Begin with ideas about associations between organism and environment; move toward concept of interdependence. ➢ Encourage student's own questions about living things and how they live. ➢ Ignore student's anthropomorphic explanations. ➢ Use hand lenses to extend the senses.
Middle School Level *(NSES pgs. 156-158; Benchmarks p. 104; SFAA pgs. 60, 61)* ➢ All organisms have instructions that specify their traits. ➢ Hereditary information passes between generations through genes and chromosomes. ➢ Plants obtain energy from the sun; animals must consume food. ➢ Animals of the same species share similarities and differences. ➢ Evolution accounts for the diversity among species. ➢ Members of a species can mate and produce fertile offspring. ➢ Diversity arises through adaptations to changing environments. ➢ Biological classification relies upon comparisons of internal and external structures. ➢ Diversity enhances the survival of a species. ➢ Many organisms that once lived are extinct.	**Middle School Level** *(NSES p. 155; Benchmarks pgs. 104, 340)* ➢ Students focus on observable traits. ➢ Students have restricted understanding of plants. ➢ Hierarchical grouping is possible. ➢ Understanding adaptation may be problematic. Believe that individuals can deliberately adapt to environmental change.	**Middle School Level** *(NSES p. 155; Benchmarks p. 104)* ➢ Move from studying individuals toward studying populations and communities. ➢ Begin to explore quantitative interactions among living things.
High School Level *(NSES pgs. 185-186; Benchmarks p. 105; SFAA pgs. 60, 61)* ➢ Species are the unit for classifying organisms. ➢ Hereditary information is coded in DNA molecules. ➢ The great diversity of species evolved over 3.5 billion years. ➢ Biological classification is based upon evolutionary relationships. ➢ DNA evidence can be used to estimate degree of relatedness among species.	**High School Level** *(NSES p. 181; Benchmarks pgs. 105, 340)* ➢ Students may still have difficulty with abstract concepts such as evolution. ➢ New variations in a population may be attributed to an organism's need or the extent of use. ➢ The association between natural selection and differential reproduction may be difficult to grasp. ➢ Student may make classification errors because of an organism's common name.	**High School Level** *(NSES p. 181; Benchmarks p. 105)* ➢ Describe evolution as a phenomenon of populations, not individuals. ➢ Emphasize the randomness of mutations. ➢ Make the association between survival and reproductive success. ➢ Successful understanding of DNA depends upon being able to understand molecules.

REFLECTION 3.1 Content Clarification of a National Standard

> "Discovery consists of seeing what everyone has seen and thinking what nobody has thought."
>
> **Albert Szent-Gyorgi**

1. After completing the Content Clarification, my understanding about the *content knowledge* associated with this national standard has:

☐ Changed ☐ Stayed the Same

2. After completing the Content Clarification, my understanding about *how students learn* the content associated with this national standard has:

☐ Changed ☐ Stayed the Same

3. After completing the Content Clarification, my understanding of *how to teach* the content associated with this national standard has:

☐ Changed ☐ Stayed the Same

4. After thinking about these three aspects of the Content Clarification, here are some future steps that I would consider when targeting this particular standard with my students:

A.

B.

C.

EXPLANATION 3.2 Creating a Content Crosswalk

One of the major premises of the Guide is that learning about standards is neither simple nor direct. This Explanation incorporates a process called a content crosswalk that should enable you to increase your understanding of specific learning goals for students.

> *"Science literacy should be approached not as a collection of isolated abilities and bits of information, but as a rich fabric of mutually supporting ideas and skills that must develop over time."*
>
> **American Association for the Advancement of Science, 2001**

The crosswalk matrix is a visual tool for comparing common themes across different sources. Many organizations, including the Council of Chief State School Officers, the Maine Mathematics and Science Alliance, Project 2061 and State Departments of Education have employed this technique to facilitate cross-comparisons between standards documents. In this content crosswalk you will search the national standards resources for direct and indirect references to the learning goal that you are investigating.

PART 1: STRUCTURE OF A CONTENT CROSSWALK TEMPLATE

These scaffolded questions are aimed at helping you to understand the structure of a Content Crosswalk. Figure 3.3 on page 36 is a sample from the State of Tennessee's Curriculum Crosswalk that was modeled after the work of the Maine Mathematics and Science Alliance (Keeley, 2001). The following questions refer to this figure. Note that Tennessee uses the expression *learning expectation* synonomously for a *learning goal.*

5. What do the headings of the remaining columns represent?	
4. What information do you find in the far left-hand column?	
3. Which grade clusters are referenced?	
2. What Tennessee content standard does the crosswalk target?	
1. What specific content area is addressed?	

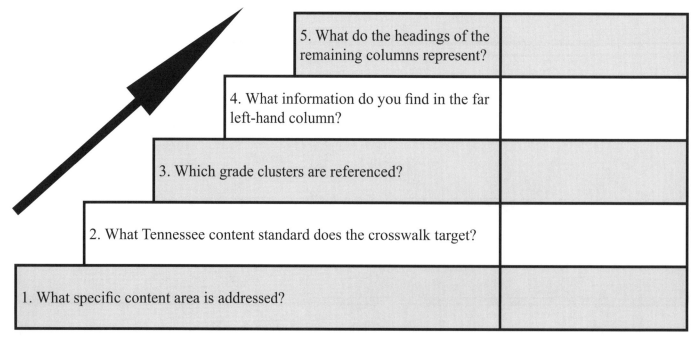

EXPLANATION 3.2 Creating a Content Crosswalk

6. What is the significance of the national standards resources to the Content Crosswalk procedure?

7. What is the connection between the national resource documents referenced in the columns and Tennessee's Learning Expectations (goals)?

8. We refer to the information found in the individual boxes as *sightings*. What is your interpretation of this term?

FIGURE 3.3 TENNESSEE CURRICULUM CROSSWALK

Earth and Space Science Standard 8.0: Atmospheric Cycles

(Changes in the atmosphere control weather and climate)

	Tennessee's Learning Expectations	*National Science Education Standards*	*Science for All Americans*	*Benchmarks for Science Literacy*	*Atlas of Science Literacy*
K-3	**8.1 - Recognize daily and seasonal weather changes.**	K-4/C3b pg. 134 K-4 Essay pgs. 131-133	The Physical Setting pg. 43	4B (K-2) / 1 Essay (K-2) pg. 67	No Map
	8.2 - Recognize that weather is associated with temperature, precipitation, and wind conditions and can be measured using tools and instruments.	K-4/C3b pg. 134 K-4 Essay pgs. 131-133	The Physical Setting pg.43	Essay pg. 67	No Map
4-5	**8.3 - Recognize that atmospheric conditions vary and can be measured.**	K-4 Essay pgs. 131-133	The Physical Setting pgs. 43-44 The Living Environment pg. 66	4B (K-2) / 1 pg.	No Map
	8.4 - Recognize that landforms and bodies of water affect weather and climate.	5-8/C1j pg. 160	The Physical Setting pgs. 42-45	4B (6-8) / 6, 7 pg.	No Map
	8.5 - Understand the basic features of the water cycle.	5-8/C1f pg. 160	The Physical Setting pg. 43	4B (3-5) / 3 Research 4B pg. 336	Conservation of Matter pg. 57 States of Matter pg. 59
6-8	**8.6 - Interpret the relationship between weather and the water cycle.**	5-8/C1l pg. 160	The Physical Setting pg. 43	4B (6-8) / 7 Research 4B pg. 336	No Map
	8.7 - Investigate the relationship between the collection of weather data and its interpretation.	K-4 / C3b pg. 134 K-4 Essay pgs. 130, 134	The Nature of Technology pg. 26 The Physical Setting pgs. 43-44 Common Themes pg. 175	4B (6-8) / 6 pg.	No Map

EXPLANATION 3.2 Creating a Content Crosswalk

PART 2: YOUR STATE OR LOCAL STANDARDS DOCUMENT

This is an opportunity for you to discover more about the standards document that is most relevant to your own classroom practice.

5. What are the different K-12 grade clusters?	
4. How many standards are found in each content area?	
3. What major content areas were included?	
2. Who was involved with preparing this document?	
1. When was your local state or local standards document prepared?	

6. What is the difference between a content standard, a benchmark, and a learning goal?

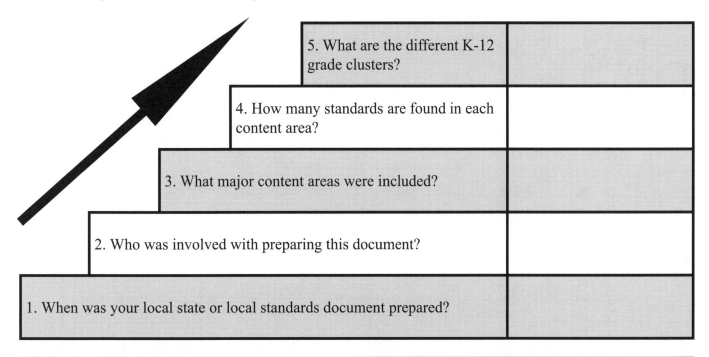

EXPLANATION 3.2 Creating a Content Crosswalk

PART 3: BUILDING A FRAMEWORK FOR YOUR OWN CONTENT CROSSWALK

Select a state or local content standard that is associated with a course or grade level that you teach. Because you are trying to gain more experience with standards, it would be better to work with a different standard from the one you investigated earlier. Repeat the first part of Explanation 3.1 if you have difficulty choosing a standard to study.

DIRECTIONS

- Write the state or local standard that you selected in the appropriate space in the Content Crosswalk template (Figures 3.4 and 3.5). Note that separate templates are provided for mathematics and science.
- Place the titles of all the national documents and your major classroom curriculum resource for the content area that you are referencing in the columns.
- List *all* of the K-12 learning goals associated with this standard in the first column of the Content Crosswalk template.
- Place stars beside the learning goals that you target in the grade level or course that you teach.
- Focus on *one* of the learning goals that you target in the course or grade level that you teach.
- Follow this one learning goal horizontally across the row. Note the titles of the national documents that you should be referencing.
- Carefully look for sightings of this learning goal in each of the national standards resource documents and the instructional materials used in your classroom. Sightings include any and all direct or indirect references to the learning goal that you are considering.
- Enter all of the sightings you make in the crosswalk template.

Congratulations! You have just constructed a small piece of a Content Crosswalk. For now, it is not important to read or analyze these sightings. You will do this immediately before building a student assessment, selecting the learning activities in your standards-based material, or evaluating a curriculum material.

FIGURE 3.4 DISCOVERY GUIDE: CONTENT CROSSWALK TEMPLATE (SCIENCE)

Standard:

Grade Levels	Learning Goals	*National Science Education Standards*	*Science for All Americans*	*Benchmarks for Science Literacy*	*Your Principal Classroom Resource*

FIGURE 3.5 DISCOVERY GUIDE: CONTENT CROSSWALK TEMPLATE (MATHEMATICS)

Standard:

Grade Levels	Learning Goals	*Principles and Standards for School Mathematics*	*Science for All Americans*	*Benchmarks for Science Literacy*	*Your Principal Classroom Resource*

REFLECTION 3.2 Creating a Content Crosswalk

1. Are the learning goals for the standard that you selected evenly or unevenly distributed among the various grade clusters? Give a plausible explanation for any distribution pattern that you find.

"What I finally understood after the crosswalk was how the learning goals related to one another...and how important it is to have an understanding of where your students are coming from and where they will be going in the years to come."

Journal entry from a past user of the guide

2. Imagine a child who has completed a grade K-12 sequence for this standard and set of learning goals. Does this collection of learning goals ensure that the student would meet the intent of the standard? Explain.

3. Do you feel that any important learning goals for your grade level were missing from your state's standards document and should be added?

REFLECTION 3.2 Creating a Content Crosswalk

4. What did you discover about the process of generating a Content Crosswalk?

5. How would you describe a Content Crosswalk to a colleague?

6. What are some possible applications of developing a full-blown Content Crosswalk in which all learning goals for all grade levels were cross-referenced to the national documents?

EXPLANATION 3.3 Applying a Content Crosswalk to Clarify a Learning Goal

In Explanation 3.2, you produced a segment of what could potentially be expanded into a full-scale Content Crosswalk. If such a document were to be produced, then *every* state or local learning goal in *all* benchmarked grade levels would be cross-referenced to their counterparts in the national standards resource documents.

> *"The dramatic changes called for in the new standards are very difficult to put into full practice and where attempted, generally fall short of the mark."*
>
> **Anderson & Helms, 2001**

The previous Explanation raised the idea that a Content Crosswalk could have important implications for making decisions about curriculum and instruction. In Explanation 3.3, we complete an in-depth analysis of the same learning goal that you selected for the previous activity. You will consider how a Content Crosswalk can create a framework for helping you to align decisions that you make about assessment, curriculum, and instruction with a targeted learning goal for students.

DIRECTIONS

1. Refer to the Content Crosswalk that you prepared in Explanation 3.2. You will be working with this same learning goal.

2. Carefully read all the cross-referenced sightings that you located in the national standards resource documents and your classroom instructional materials.

3. Place the information you have gathered in the Discovery Guide for a Content Clarification of this learning goal (Figure 3.6). Refer back to Explanation 3.1 if you experience any difficulty with this task.

   NOTE: An additional Ways to Assess Student Learning column has been added to begin the thinking process that will eventually culminate in a comprehensive assessment plan for your standards-based material.

4. After reviewing your Discovery Guide, you should complete the Summary Reflection for Chapter 3.

FIGURE 3.6 DISCOVERY GUIDE: CONTENT CLARIFICATION OF A LEARNING GOAL

State Content Standard:

State Learning Goal:

RELATED CONTENT KNOWLEDGE	DISCOVERIES ABOUT STUDENT LEARNING	IDEAS ABOUT TEACHING TO THIS LEARNING GOAL	POSSIBLE WAYS TO ASSESS STUDENT LEARNING

SUMMARY REFLECTION

CHAPTER 3

1. What did completing a Content Clarification accomplish for you?

"One's mind, once stretched by a new idea, never regains its original dimensions."

Oliver Wendell Holmes

2. What part of the Content Clarification process did you find most valuable? Least important?

3. Some educators believe that standards and learning goals lack specificity and provide little guidance for teachers. How can the national standards resources help teachers to address these perceived shortcomings?

4. Some users of the Guide have observed that preparing a Content Clarification more closely resembles pre-planning than direct planning for instruction. What are your thoughts concerning this observation?

5. If the state or local standards and learning goals that you reference in your teaching are the end points of a student's journey to learning, how can a Content Clarification provide a road map to reach these destinations?

EXPLANATION II: Assessment in a Standards-Based System

CHAPTER 4

In Chapter 4, the focus of the Guide shifts to assessment. When educational historians examine the education reform movement of the 1990s, they are apt to notice three major arenas of change. The most obvious is the influence of technology. Students and teachers gained unprecedented access to computers and to the vast information storehouse on the Internet. The second major change, in fact, the reason why this Guide was written, is the impact of standards across the full educational spectrum. The final area is associated with transformations in the way that student performance was evaluated. With respect to assessment, the emphasis shifted from approaches that were primarily geared to assigning grades, toward techniques aimed at providing clear indications of what students were actually learning.

> *"New definitions and structures for...curriculum and teaching strategies are difficult to attain. However, change in assessment strategies has been even more difficult."*
>
> **Hammerman & Musial, 1995**

EXPLANATION II

LEARNING GOALS	ACTIVITIES
Investigate connections among learning goals for students, assessment, and the instructional practices of teachers.	4.1: Congruence With Standards
Examine prior understanding of standards-based assessment.	4.2: Traditional Versus Standards-Based Instruction
Develop a high level of familiarity with terminology associated with assessment.	4.3: The Language of Assessment
Identify the essential features of a performance assessment.	4.4: Elements of a Performance Assessment
Design a performance assessment that targets a particular learning goal.	4.5: Building Your Performance Assessment
	4.6: Rubrics for Scoring Student Work

SUMMARY READINGS AND REFLECTIONS

1. *Facing and Embracing the Assessment Challenge* (Damian, 2000)

2. *The Authentic Standards Movement and Its Evil Twin* (Thompson, 2001)

GUIDED REFLECTION

CHAPTER 4

At this point, it might be helpful to reexamine what you have hopefully gained from the Guide and to further reflect upon the standards-based world that you are entering. We'll do this through the eyes of a fictional character, Julian, a seventh-grade teacher from a small middle school in Tennessee, and let him represent a typical user of the Guide.

Three things originally prompted Julian to invest his time and energy in the Guide. First, he's become a strong believer in standards-based reform, but has little direct experience translating these recommendations for change into teaching and learning. Second, Tennessee just completed a full revision of its state science standards, standards that he'll be responsible for implementing in his classroom. Finally, Julian doesn't receive many opportunities for formal professional development.

Here's an account of his experience thus far with the Guide…

In *Chapter 1*, Julian explored his personal beliefs about change and his prior understanding of standards and constructivist teaching. He discovered that he has a positive attitude toward change...one that could certainly support an effort to modify his teaching if it served the best interests of his students. From his survey results on standards, Julian concluded that he currently held, at best, a sketchy understanding of what these ideas were all about.

Completing *Chapter 2* was a revelation for Julian. Until now, he really hadn't been aware of the extensive resources that were available to help him gain a better understanding of the standards-reform movement. Julian was surprised and tremendously reassured when he investigated these documents. The national materials provided the background information about standards that he needed to support and justify the types of changes he envisioned for his own classroom.

Until *Chapter 3*, few of Julian's experiences with the Guide seemed to be connected with issues that had a direct bearing on his classroom or with the new state science standards. Two important events happened during the Explanation that transformed his ideas about teaching to standards. First, Julian was introduced to the Content Crosswalk. This gave him a systematic procedure for referencing the new Tennessee standards to their counterparts in the national standards resources. Figure 4.1 illustrates the standard that Julian investigated and the sightings from his crosswalk.

Second, after building the crosswalk, Julian applied a Content Clarification to research the ideas associated with one of the learning goals that he is responsible for targeting in his course. Figure 4.2 illustrates what Julian discovered through his Content Clarification. As he unearthed information concerning content knowledge, research-based data on student learning, and concerns about instruction, a full picture of the underlying intent behind Learning Goal 7.5 materialized. After completing this investigation, Julian felt confident and ready for the task of planning instruction that would effectively target this learning goal. However, Julian still had lots of important work ahead, especially in the areas of assessment and standards-based instructional design.

FIGURE 4.1 DISCOVERY GUIDE: JULIAN'S CONTENT CROSSWALK

Earth and Space Science Standard 7.0: Earth and Its Place in the Universe

(The universe is composed of many galaxies made up of solar systems in continuous motion. This motion is governed by the force of gravity.)

Grade	Tennessee's Learning Expectations	National Science Education Standards	Science for All Americans	Benchmarks for Science Literacy	Atlas of Science Literacy
K-3	**7.1 - Recognize that different objects appear in the day and nighttime sky.**	K-4 Essay pg. 130	The Physical Setting pgs. 40-41	4A (K-2) / 2 Essay (K-2) pg. 62	Solar System pg. 45 Stars pg. 47
K-3	**7.2 - Recognize that there are predictable patterns that occur in the universe.**	K-4/C3c pg. 134	The Physical Setting pg. 43	4A (K-2) / 3 Essay (K-2) pg. 62 Research 4B pgs. 335-336	Solar System pg. 45
4-5	**7.3 - Know that objects in space have identifiable characteristics, such as appearance, location, and apparent motion.**	K-4/C2a pg. 134	The Physical Setting pg. 41	4A (3-5) / 1	Gravity pg. 43 Solar System pg. 45 Stars pg. 47
4-5	**7.4 - Investigate the patterns and movement of objects in space.**	K-4/C3a pg. 134	The Physical Setting pgs. 40-41	4A (3-5) / 1 Essay (3-5) pg. 62	Gravity pg. 43 Solar System pg. 45 Stars pg. 47
6-8	**7.5 - Recognize the basic features of the universe.**	5-8/C3a pg. 160 5-8/C3c pg. 161	The Physical Setting pgs. 40-41	4A (6-8) / 1, 3	Solar System pg. 45 Stars pg. 47 Galaxies and the Universe pg. 49
6-8	**7.6 - Investigate the relative distances of objects in space.**	5-8 Essay pgs. 158-159	The Physical Setting pg. 40	4A (6-8) / 2	Stars pg. 47 Galaxies and the Universe pg. 49
6-8	**7.7 - Describe the positional relationships among the earth, moon, and sun.**	5-8/C3a pg. 160	The Physical Setting pg. 41	4A (3-5) / 4	Gravity pg. 43 Solar System pg. 45
6-8	**7.8 - Understand that gravity is the force that keeps planets in orbit around the sun and governs movement in the solar system.**	5-8/C3c pg. 161	The Physical Setting pg. 42	4B (6-8) / 3 4G (6-8) / 2 Essay (6-8) pg. 68 Research 4G pg. 340	Gravity pg. 43
6-8	**7.9 - Explore the role of technology and careers associated with the study of space.**	K-4/C2d pg. 138	No Map	3A (3-5) / 4	No Map

FIGURE 4.2 DISCOVERY GUIDE: JULIAN'S CONTENT CLARIFICATION OF A LEARNING EXPECTATION

Content Standard:	**Earth and Space Standard 7.0 - Earth and Its Place in the Universe**		
Learning Expectation:	**7.5 - Recognize the basic features of the universe.**		
RELATED CONTENT KNOWLEDGE (K-8)	**DISCOVERIES ABOUT STUDENT LEARNING**	**IDEAS ABOUT TEACHING TO THIS LEARNING EXPECTATION**	**WAYS TO ASSESS STUDENT LEARNING**
SFAA: pgs. 40, 41. ➢ Our galaxy, the Milky Way, contains a billion stars of which our sun is one. ➢ Light from our nearest star takes four years to reach earth. ➢ All galaxies appear to obey the same physical principles. ➢ Our sun is the central and largest body in the solar system. ➢ Our solar system is about five billion years old. ➢ Most objects in solar system move in stable patterns. ➢ The earth is one of nine planets that orbit the sun. ➢ Different planets have different features. ➢ Tools like radio and x-ray telescopes are needed to observe the universe. ➢ Most of what we know about the universe is inferred.	**Elementary Grade Level** NSES p. 130; Benchmarks p. 62. ➢ Few children can understand the concept of a billion. ➢ Sun-centered system conflicts with earth-centered view. **Middle School Level** NSES pgs. 160, 161; Benchmarks p. 63. ➢ Because there is no direct contact with these topics, some should be reserved for later in middle school. ➢ Students' grasp of understanding about enormous distances and large time scales must grow slowly over time.	**Elementary Grade Level** NSES p. 130; Benchmarks p. 62. ➢ Children should be *guided* to observe changes and patterns in the day and nighttime sky. ➢ Instruction should be limited to making and describing observations. ➢ Learning the names of constellations is unimportant. **Middle School Level** NSES p. 158; Benchmarks p. 62. ➢ Idea of systems is critical to understanding this topic. ➢ Direct experimentation is difficult with this topic. ➢ Computer simulations and scale models are the most effective learning activities.	**Middle School Level** ➢ Pre-assessment: draw an illustration showing objects in the sky and their relative movements. ➢ Daily sky watching; graphing of observations. ➢ Use toilet tissue to create a relative distance model for the planets. ➢ Create a travel brochure for a planet. ➢ Complete a WebQuest for this topic.

EXPLANATION 4.1 Congruence With Standards

The central questions addressed by the principle of congruence are (a) How does a teacher take a learning goal and use it to design an assessment that provides valid and sufficient evidence that this goal has been achieved by students? and (b) How does a teacher then use this assessment to guide his or her selection of learning experiences that enable students to demonstrate that they have attained this learning goal?

> *"As with other design professions, standards inform and shape our work...We are not free to teach any topic we choose. Rather, we are guided by national, state, district, or institutional standards that specify what a student should know or be able to do."*
>
> **Wiggins & McTighe, 1998**

In Explanation 4.1, you complete a careful analysis of the congruence triangle, a graphic organizer that provides the conceptual foundation for this chapter of the Guide. This model depicts the intertwining of principles and ideas associated with standards-based reform (Reynolds, Doran, Allers, & Agruso, 1996). Visual representations of information can be powerful tools for understanding because of their ability to represent complex patterns in meaningful ways (Hyerle, 1996). We believe that the congruence triangle has the capacity to clarify the central issues of curriculum reform.

Although standards are the *ends* for student learning, they represent the starting points for assessing student understanding and designing instruction. The principles of congruence provide an organizational framework for developing standards-based assessments and instructional materials. The following set of questions directs you through an analysis of the main ideas associated with curriculum congruence.

1. Carefully examine the congruence triangle (Figure 4.3 on page 53). What are three thoughts that arise from your study of this illustration?

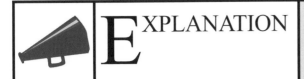

EXPLANATION 4.1 Congruence With Standards

2. What do you think that the terms *validity, correlation,* and *alignment* mean within the context of the Reynolds model?

3. In the Reynolds model, there is no recommended starting point for planning instruction. If you had to select from the three alternatives, where would you begin and why?

4. Clearly and concisely explain your present understanding of the expression "curriculum congruence."

5. Do you think it is important for standards, assessment, and instruction to be fully congruent? Why, or why not?

6. What is the relevance of the principle of congruence to someone like yourself who has embarked on a mission to develop standards-based materials?

7. Do conceptual models like the congruence triangle help you to gain a better understanding of complex ideas, and if so, how?

FIGURE 4.3 CONGRUENCE TRIANGLE

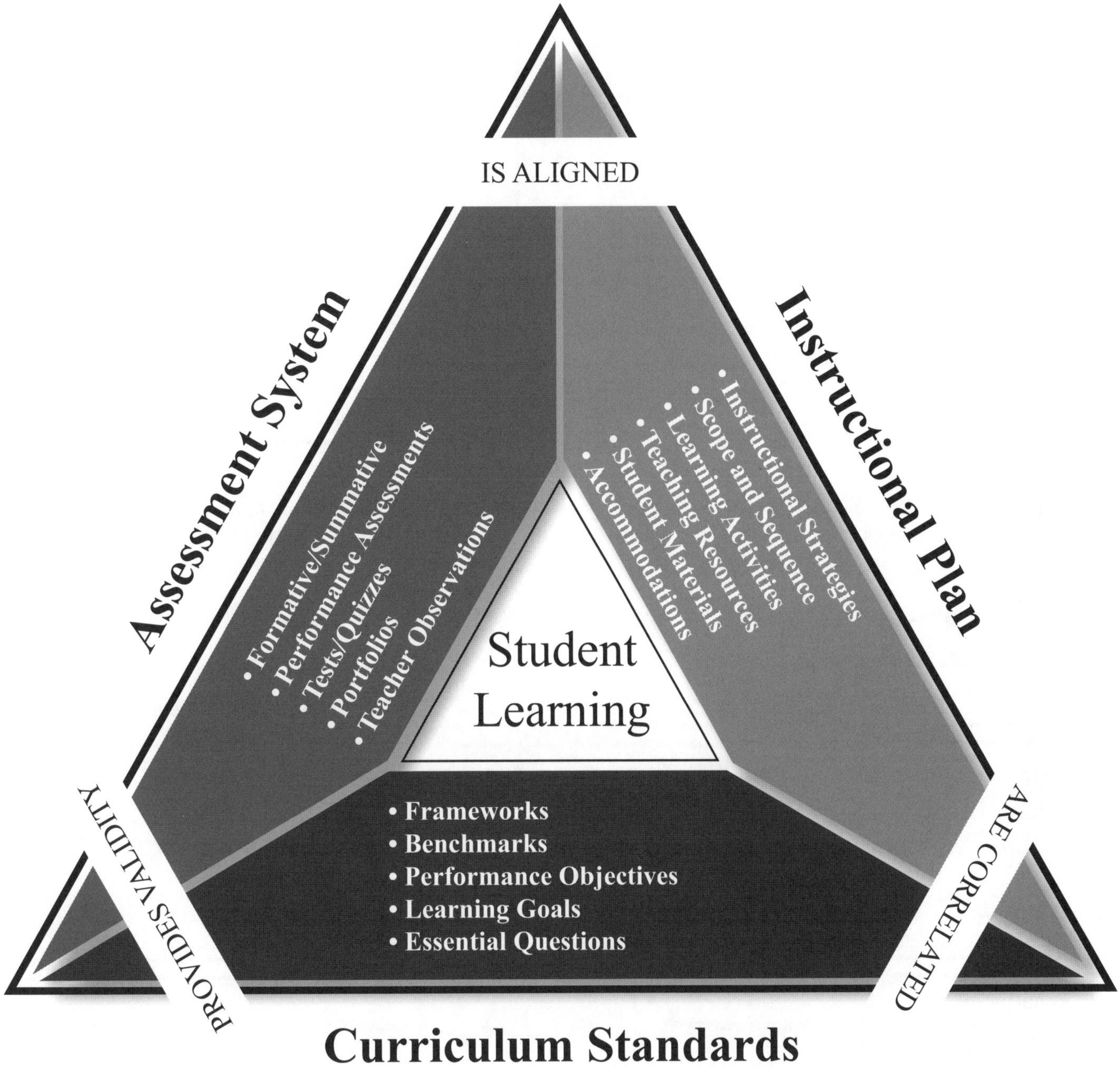

Adapted from Reynolds, D., Doran R., Allers, R., & Agruso, S. (1996). *Alternative assessments: A teacher's guide*. Buffalo, NY: University of Buffalo Press.

EXPLANATION 4.2 Traditional Versus Standards-Based Instruction

> *"Standards are descriptions of what ... instruction should enable students to know and do."*
>
> **Naional Council of Teachers of Mathematics, 2000**

One important element of teaching expertise is the ability to skillfully construct comprehensive instructional materials that offer genuine opportunities for students to learn. In this Explanation, you investigate the essential features of a standards-based model for designing curriculum and contrast this method with more typical approaches. You may want to visit the *Understanding by Design* Web site (**www.ubdexchange.org**) to discover why the method described in this activity is commonly referred to as 'backward design (Wiggins & McTighe, 1998).

DIRECTIONS

Figure 4.4 illustrates two contrasting images of the instructional planning process. After studying this diagram, you should complete the Venn diagram in Figure 4.5 to analyze the similarities and differences between these approaches for selecting and organizing learning experiences for students.

FIGURE 4.4 INSTRUCTIONAL PLANNING APPROACHES

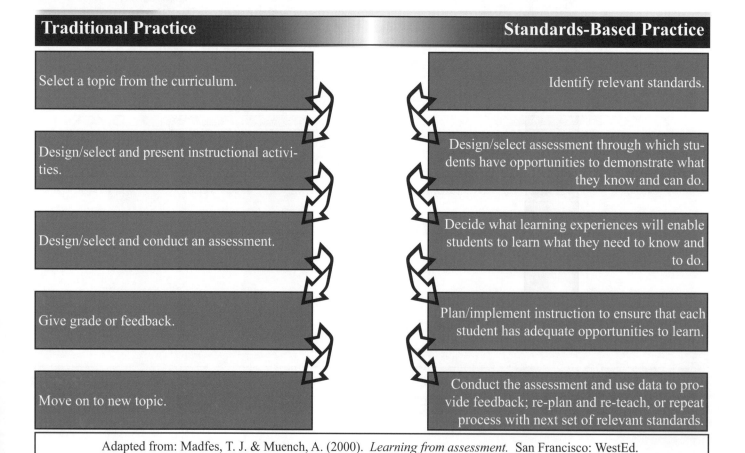

Traditional Practice	Standards-Based Practice
Select a topic from the curriculum.	Identify relevant standards.
Design/select and present instructional activities.	Design/select assessment through which students have opportunities to demonstrate what they know and can do.
Design/select and conduct an assessment.	Decide what learning experiences will enable students to learn what they need to know and to do.
Give grade or feedback.	Plan/implement instruction to ensure that each student has adequate opportunities to learn.
Move on to new topic.	Conduct the assessment and use data to provide feedback; re-plan and re-teach, or repeat process with next set of relevant standards.

Adapted from: Madfes, T. J. & Muench, A. (2000). *Learning from assessment.* San Francisco: WestEd.

FIGURE 4.5 INSTRUCTIONAL PLANNING APPROACHES

Traditional Practice *Commonalities* **Standards-Based Practice**

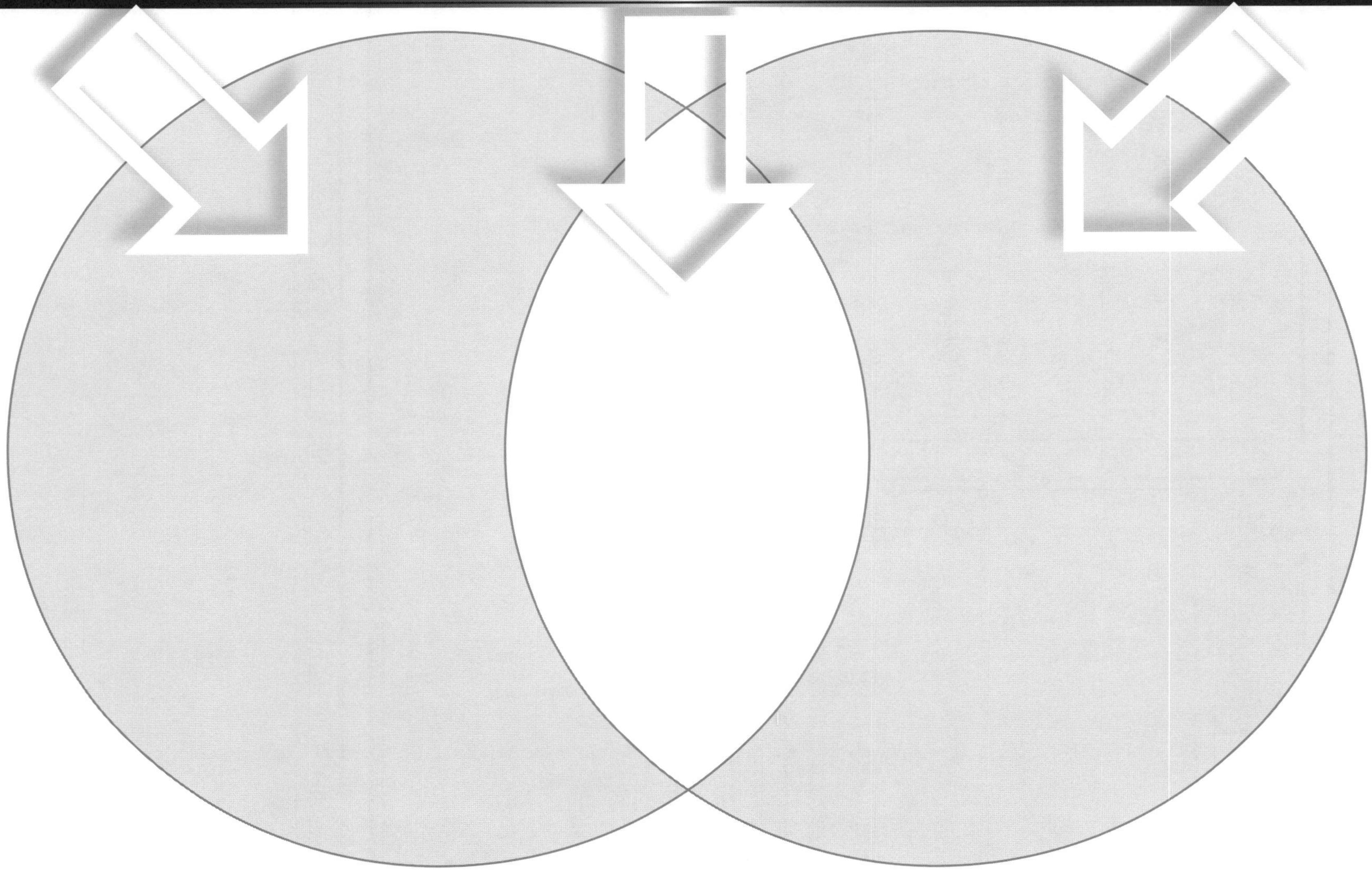

Summary of Findings:

REFLECTION 4.2 Traditional Versus Standards-Based Instruction

1. What do you regard as the major difference between traditional and standards-based approaches to designing instruction?

"The more I use standards, the easier they become. As this class goes on I am finding that there is no substitute for (first) familiarizing yourself with the standards…I am seeing the wisdom of this method."

Clark, 2001

2. One of the distinctive features of standards-based practice is the placement of the assessment component. Do you believe that it is feasible for a teacher to develop an effective assessment plan before selecting the instructional activities? Explain.

3. Which of the two models makes the most sense to you? What makes this approach preferable?

REFLECTION 4.2 Traditional Versus Standards-Based Instruction

4. What do you see as the most immediate difficulty in implementing a standards-based model?

5. Do you believe that the designation of the first model as "traditional" connotes or suggests that it is a less desirable approach? Explain your answer.

6. One popular name for the standards-based instructional design model is "backward design." What is your interpretation of this expression?

EXPLANATION 4.3 The Language of Assessment

In all professions, the core language changes over time. New words creep into the vocabulary and older ones take on new meanings. Other expressions lose their utility and eventually disappear.

> *"Formative assessment is when the chef tastes the soup; summative assessment is when the customer tastes the soup."*
>
> **Author Unknown**

Educators often are faulted for using what some regard as needless jargon. Yet one of the characteristics of every professional community is that it possesses a unique language that its members rely upon to support and enhance communication. Jargon is most useful when it conveys meaning precisely. Such terms and expressions constitute the lexicon of our distinctive profession of teaching. Nowhere is this explosion of new language more apparent than for the subject of assessment. In this Explanation, we investigate terms that are currently overheard in conversations among professionals about this important topic.

DIRECTIONS

1. Make a photocopy of Figure 4.6.

2. Carefully fold the sheet along the dashed line *before* you examine Figure 4.6.

3. Step 1 involves reviewing terms commonly used in discussions about assessment. You should enter your own working definitions in the spaces provided in Figure 4.6. The key word to note here is "working." These are your personal operational definitions and need not be perfectly refined constructions.

4. After you insert your working definitions, open the sheet fully.

5. Now, carefully compare your own descriptions with a formal textbook definition.

6. After you make these comparisons, enter any observations, insights, and conclusions in the comment spaces provided in Figure 4.6. To wrap up your study of the Language of Assessment, you should complete the follow-up reflection for this section.

FIGURE 4.6 THE LANGUAGE OF ASSESSMENT

Term	My Working Definition	Textbook Definition	Comments
Alternative Assessment		Any assessment that is not of the multiple-choice, matching, and true-false, paper-and-pencil formats. Requires students to complete a task or demonstrate a performance.	
Analytic Scoring		A scoring procedure in which performances are evaluated for selected indicators, with each indicator receiving a separate score.	
Assessment		The systematic collection of information about student achievement and performance.	
Authentic Assessment		Assessment tasks that use "real-world" and "real life" contexts, and are aligned with the assessment and content standards in use by your school or district.	
Content Standards		Goal statements identifying the knowledge, skills, and dispositions to be developed through instruction in the content areas.	
Evaluation		The interpretation of assessment data, based on a set of criteria, to judge student achievement and capabilities.	
Formative Assessment		The collection of data about student achievement and performance that is part of regular teaching and learning in classrooms.	
Holistic Scoring		A scoring procedure yielding a single score based upon an overall impression of a product of performance.	
Indicator		Provides a more specific description of an outcome in terms of observable and assessable behaviors.	
Peer Evaluation		When other students or peers score a student's assessment in relation to the established criteria for success.	
Performance Assessment		An assessment activity that requires students to construct a response, create a product, or perform a demonstration.	
Portfolio		Collection of student work that can include assignments, projects, tests, or any other samples of work that illustrate the development of skills and conceptual understanding in a domain of knowledge or skills.	
Rubric		An established set of criteria for scoring or rating student assessment tasks where the response is more involved than selecting an answer from a prescribed list.	
Self-Assessment		The process where a student judges his or her achievement in relation to personal goals.	
Standardized Test		An assessment that is administered and scored the same way for all students to enable comparison of scores.	
Summative Assessment		Assessments given at the end of a unit, semester, or course where the data are used to generate grades and marks.	
Test		An assessment task, primarily for measuring student achievement in a domain of knowledge or skills, that is summative in nature.	
Validity		The extent to which an assessment measures what it sets out to measure. An assessment can be reliable but not valid.	

REFLECTION 4.3 The Language of Assessment

1. On a scale of 1 to 10, how would you rate your initial understanding of contemporary language associated with assessment? Using the same scale, how would you grade yourself after completing this Explanation?

> *"In the new view, assessment and learning are two sides of the same coin. When student engage in assessments, they should learn from those assessments."*
>
> **National Research Council, 1999**

2. What three conclusions did you reach after completing the assessment foldout activity?

3. What, if any, reservations do you have with traditional approaches to evaluating student achievement that rely principally on end-of-unit or chapter tests?

4. Use the Internet or print resources to explore the fundamental differences between *formative* and *summative* assessments. What are some specific examples of each type that you currently use with students?

REFLECTION 4.3 The Language of Assessment

5. Proponents of revamping how we assess students use expressions like "comprehensive assessment systems" and "multifaceted approaches" to describe new and preferred methods. How do you interpret these expressions, and do you think they are realistic to implement?

6. Why do you think that there is currently such a strong emphasis being placed on student assessment?

7. What is the connection between standards-based reform and the heightened importance of new models for assessing student performance?

8. What would you identify as your own most pressing needs regarding assessment? What are your plans for addressing these issues?

EXPLANATION 4.4 Anatomy of a Performance Assessment

PART A

What are your plans for assessing what you ultimately gain from your experience with the Guide? Whether you are using this book as a member of a professional development team or individually, you probably established some personal goals for improvement. It is a fair assumption that your original action plan was connected with implementing standards-based instruction. Ultimately, then, the best measure of success may lie in your ability to translate principles underscored by the Guide to create curriculum materials of your own design, or to evaluate existing curricula. This is the direction in which the Guide is headed.

> *"Performance assessments are more challenging than traditional assessments. They require more work by the teacher to create and far more work by the student to complete. Nevertheless, students frequently prefer these assignments because they are realistic, engaging, and most importantly, do not involve the 'one-shot' terrorism of the traditional test."*
>
> **Reeves, 1998**

In the approach for designing standards-based materials presented in the Guide, the first action is to target and then clarify a content standard and learning goal (Chapter 3). The next step is for teachers to build a performance assessment that is aligned with this standard and clearly demonstrates what students should know or be able to do. In a standards-based classroom, the summative or culminating assessment is introduced prior to instruction so that students gain an unambiguous picture of expectations before beginning work on new material. For this reason, knowing how to create a performance assessment is absolutely necessary for moving forward in the Guide.

DIRECTIONS

1. Figure 4.7 illustrates the three essential features of a performance assessment.

2. Take sufficient time to carefully study this diagram so that you develop a full understanding of its features. To be most effective, every performance assessment should include all three elements.

FIGURE 4.7 ESSENTIAL FEATURES OF A PERFORMANCE ASSESSMENT

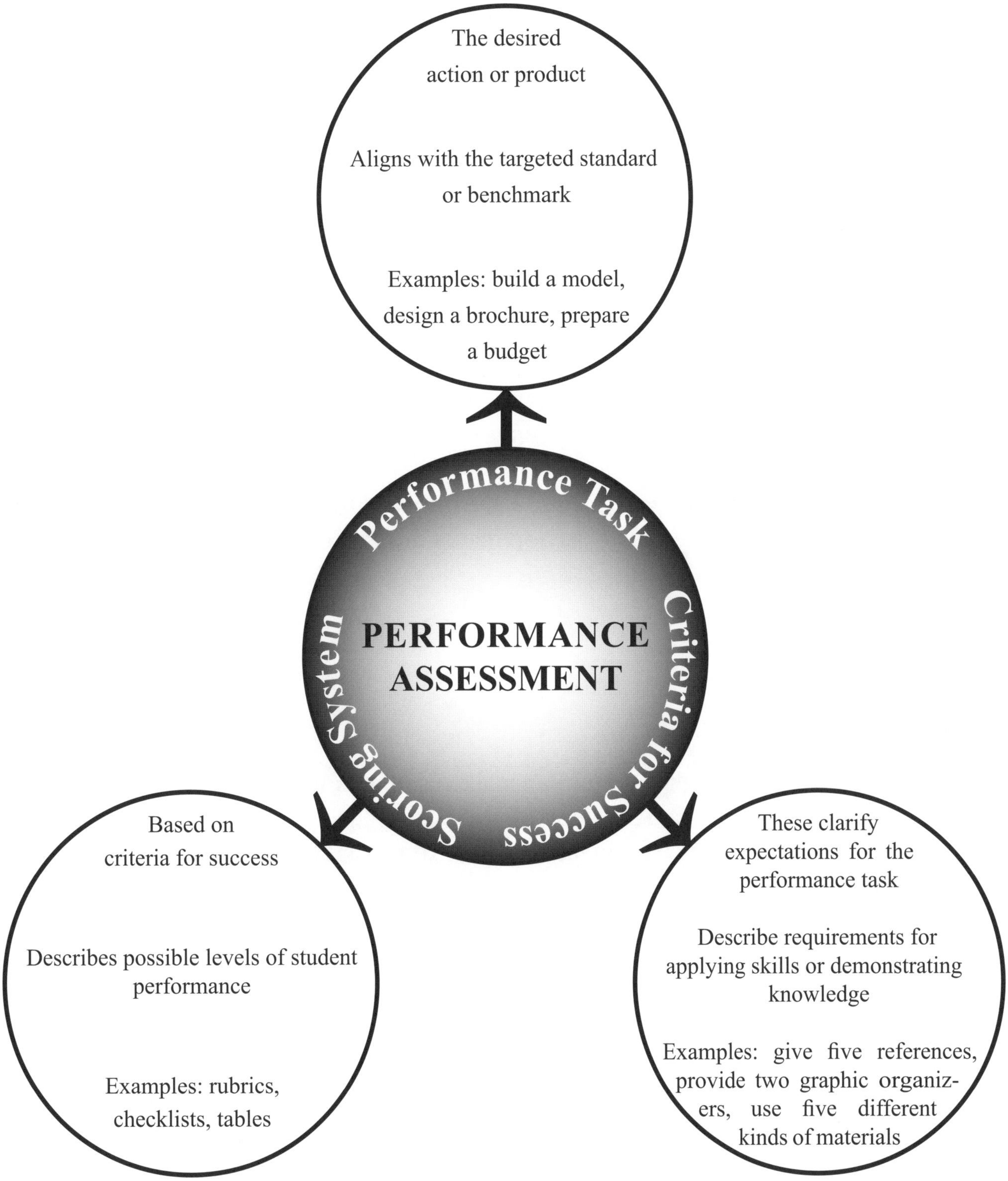

Adapted from Brown, J. H. & Shavelson, R. J. (1996). *Assessing hands-on science*. Thousand Oaks, CA: Corwin.

EXPLANATION 4.4 Analysis of a Sample Performance Assessment

PART B
DIRECTIONS

1. Together, the sample performance task on the following page and Figure 4.8 on page 66 constitute a performance assessment that was actually used with teachers to evaluate their work in a professional development course on standards-based instruction. We present this assessment here only as an example. *You should not be concerned if some of the language is new or unfamiliar.*

2. The following questions relate to the performance task and Figure 4.8. Some may be challenging. Use the information found in Figure 4.7 on page 63 for assistance.

- As in all performance assessments, a specific *performance task* must be completed. What is the expected product or action that is described?

- The second element of a performance assessment is the *criteria for success*. These define the expected characteristics of the performance task. In our example, what specific evidence must be included to demonstrate that the task has been successfully completed?

- The final part of a performance assessment is the *scoring guide*. In the example that we have been studying, the rubric shown in Figure 4.8 is aligned with the above task and also addresses the criteria for success. Study the rubric carefully. What is the relationship among the features of the rubric shown in Figure 4.8, the performance task, and the criteria for success?

EXPLANATION 4.4 Discovery Guide: *Sample Performance Task*

With guidance and support from the new Director of Curriculum and Instruction, your school district recently embarked on an ambitious program to implement standards and benchmarks across the K-12 curriculum. This is just the type of leadership and support that you've been waiting for!

To provide the necessary professional development assistance, the Director purchased copies of the Guide for every teacher in the system. Her major goal is to prepare teachers for implementing standards-based instruction. Most teachers are working in small study groups; some are working alone. Everyone is involved and seemingly eager to explore standards-based reform issues as they affect student performance in the classroom.

A second purpose of this initiative is to develop a digital library of high-quality, standards-based materials. This is because one factor that seems to hinder the implementation of standards is the scarcity of exemplary standards-based curricula. The Director believes that professional development that combines the principles embedded in the Guide with the availability of model instructional materials will establish an excellent framework for moving this district forward into a standards-based world.

Your Stake in the Situation

To receive a stipend for participating in this professional development experience, teachers are required to submit their standards-based curriculum materials for inclusion in the digital library collection. To ensure consistency, the Director established a set of guidelines for the curriculum materials drawn from the principles embedded in the Guide. Every instructional piece should show

- Careful and close congruence among the selected learning goals for students, the performance assessment, and the instructional plan
- Significant evidence that the unit was developed according to backward design principles
- An authentic performance assessment that includes all of the essential components
- An instructional plan that is aligned with the assessment and the targeted learning goals for students

Everyone's contribution to the district's Digital Library of Standards-Based Materials is important! Compliance with the Director's guidelines requires that before submitting your work, you apply the self-assessment scoring rubric found in Figure 4.8.

FIGURE 4.8 SAMPLE SCORING GUIDE

	Exceeds the Standard	Meets the Standard	Moving Toward the Standard	Below the Standard
Congruence	• Unit is fully aligned with standards and learning goals. • Assessment is fully linked to standards and learning goals. • Instructional plan is fully matched to standards and learning goals.	• Unit is generally aligned with standards and learning goals. • Assessment is generally linked to standards and learning goals. • Instructional plan is generally matched to standards and learning goals.	• Unit is partially aligned with standards and learning goals. • Assessment is partially linked to standards and learning goals. • Instructional plan is partially matched to standards and learning goals.	• Unit is not aligned with standards and learning goals. • Assessment is not linked to standards and learning goals. • Instructional plan is not matched to standards and learning goals.
Curriculum Design	• Clear evidence of backward design. • Standards and learning goals clearly identified.	• Some evidence of backward design. • Standards and learning goals identified.	• Little evidence of backward design. • Standards and learning goals poorly referenced.	• No evidence of backward design. • Standards and learning goals not identified.
Performance Assessment	• Clear and concise performance task. • Highly appropriate criteria for success. • Valid scoring system.	• Clear performance task. • Appropriate criteria for success. • Scoring system generally valid.	• Unclear performance task. • Unclear criteria for success. • Scoring system somewhat valid.	• No performance task. • Inappropriate criteria for success. • Invalid scoring system.
Instructional Plan	• Developmentally appropriate. • Promotes higher-level thinking. • Uses a broad variety of materials and resources. • Extremely logical scope and sequence. • Extremely engaging learning activities.	• Developmentally appropriate. • Promotes some higher-level thinking. • Uses a variety of materials and resources. • Logical scope and sequence. • Engaging learning activities.	• Developmentally inappropriate. • Promotes little higher-level thinking. • Little variety in materials and resources. • Poorly organized scope and sequence. • Mildly engaging learning activities.	• No attention to developmental concerns. • Promotes no higher-level thinking. • No variety in materials and resources. • Illogical scope and sequence. • Unengaging learning activities.

EXPLANATION 4.5 Building Your Performance Assessment

Explanation 4.5 addresses a question posed earlier when the congruence triangle was being examined: how does a teacher take a learning goal and use it to design an assessment that provides valid and sufficient evidence that this goal has been attained by students? In this section, you will construct a performance assessment consisting of two of the three major parts illustrated in Figure 4.7 on page 63.

> *"Leaders of the authentic assessment movement cheerfully accept the fact that teachers 'teach to the test.' Schools, they argue, should teach to tests...tests that serve our (and our students') educational goals."*
>
> **Hart, 1994**

From this point forward, all of your instructional design work will be aligned with the learning goals that you are targeting for your students. Once you make a firm decision to work with the end in mind, then every element of instruction that follows—your choice of assessment, the scope and sequence of your curriculum, and your instructional patterns—must all be internally consistent.

PART A: PERFORMANCE TASK AND CRITERIA FOR SUCCESS

When a student completes a performance assessment at the end of a curriculum unit, the result is some product or action that yields clear evidence about what he or she knows or is able to do with respect to specific learning goals. Performance assessments can be designed for individuals or shared by groups of students.

DIRECTIONS

The framework pictured in Figure 4.9 on page 68 describes the typical design elements of a *performance task* along with the appropriate *criteria for success*. Please review this diagram carefully. This is the format that we suggest using when you design the following two essential components of a performance assessment. Later, during Explanation 4.6, you will prepare a scoring rubric as the final piece of your three-part performance assessment.

FIGURE 4.9 DESIGN DETAILS FOR A PERFORMANCE TASK

Performance Task and Criteria for Success

element

Role or Persona…

- The character that a student or group of students is asked to assume.
- *Sounds like*: You are a…
- *Acts like*: explorer, team of scientists, television producer, marketing director.

element

Targeted Behavior…

- Specific behavior addressed in the performance assessment.
- *Sounds like*: Who has been asked to…
- *Acts like*: investigate a crime, organize a pressed flower collection, serve on a selection committee…

element

Audience…

- Person or group for whom the task is being performed.
- *Sounds like*: Who has been asked by…
- *Looks like*: panel of judges, principal of the school, magazine editor, mayor of a town…

element

Product or Performance…

- The precise object or act that students are expected to create or perform.
- *Sounds like*: Who will…
- *Looks like*: short story, bumper sticker, recipe, résumé, flowchart…

element

Criteria for Success…

- Detailed description of the characteristics required in the product or performance.
- *Sounds like*: The product or performance should…
- *Looks like*: three-page pamphlet. Describes the life and times of a civil rights figure who made a significant contribution to the advancement of African-Americans. Include maps and a four-generation genealogical tree.

EXPLANATION 4.5 Building Your Performance Assessment

PART B: BUILDING YOUR PERFORMANCE TASK

Achieving success with the next activity is essential for completing your standards-based material. You should focus on the same learning goal for which you completed the Content Clarification in Explanation 3.3. For each segment of the task, you have options that you can incorporate into your assessment. The choices that you select depend upon the learning goals that you are targeting, the amount of intended authenticity, and your desire to motivate students.

Build your performance assessment so that it includes each of the elements listed below from Figure 4.9. You can combine these in any order provided that the text for students makes sense and is clear, user-friendly, and engaging.

- Targeted Standards
- Portrayal of the Role or Persona
- Account of the Targeted Behavior
- Description of the Audience
- Explanation of the Product or Action
- List of Criteria for a Successful Product or Action

It is reasonable for you to experience some difficulty in building this assessment. Developing this skill takes practice, so do not be discouraged. In Explanation 4.6, you will design a scoring rubric to accompany this performance task. This completed performance assessment will eventually become your guide for selecting the set of student activities and instructional strategies that addresses the learning goal that you are targeting.

Figure 4.10 gives an example of a performance task that includes all of the features listed above. You will also note that the targeted benchmarks and national science standards have been referenced.

FIGURE 4.10 SAMPLE PERFORMANCE TASK: *THE WRIGHT STUFF!*

Targeted Science Standards and Benchmarks
Benchmarks for Science Literacy—4F Motion
By the end of 5th grade, students should know that
Changes in speed or direction of motion are caused by forces. The greater the force is, the greater the change in motion will be. The more massive an object is, the less effect a given force will have.
How fast things move differs greatly. Some things are so slow that their journey takes a long time; others move too fast for people to even see them.
National Science Education Standard: Motions and Forces 5-8
The motion of an object can be described by its position, direction of motion, and speed. That motion can be measured and represented on a graph.
An object that is not being subjected to a force will continue to move at a constant speed and in a straight line.
If more than one force acts on an object along a straight line, then the forces will reinforce or cancel one another, depending on their direction and magnitude. Unbalanced forces will cause changes in the speed or direction of an object's motion.

Announcer...

Welcome future aeronauts **(ROLE)** *to Kitty Hawk, North Carolina, Birthplace of American Aviation! The Fly-By-Night Aircraft Corporation, proud sponsors of the Bernoulli Olympics, is eager to meet members of your state's winning middle school team and see the original plane design that you'll be showcasing at this year's competition* **(TARGETED BEHAVIOR)**.

Fly-By-Night organizes this annual event for young people from across the country to spark interest in plane design and scientific principles of flight. Awards include scholarships, model airplanes, and souvenir T-shirts. The national winners receive an all-expense-paid trip to the Smithsonian Air and Space Museum in Washington, D.C.

This year's judging panel consists entirely of descendants of the Wright Brothers **(AUDIENCE)**. *Bernoulli contest rules state that your plane* **(PRODUCT)** *can be made only of wood, paper, and glue; have no moving parts; and weigh no more than two pounds.*

Two days from now, on Fearless Flyer Day, you'll have three tries at flying your team's plane, and your team will be interviewed. Best flight wins! During the fly-off, judges will be looking for **(CRITERIA FOR SUCCESS)**:

- ★ *How long your plane remains in the air*
- ★ *Distance traveled*
- ★ *Accuracy of flight path*
- ★ *Appearance of plane*
- ★ *Adherence to contest rules*
- ★ *Explanation of principles of flight*

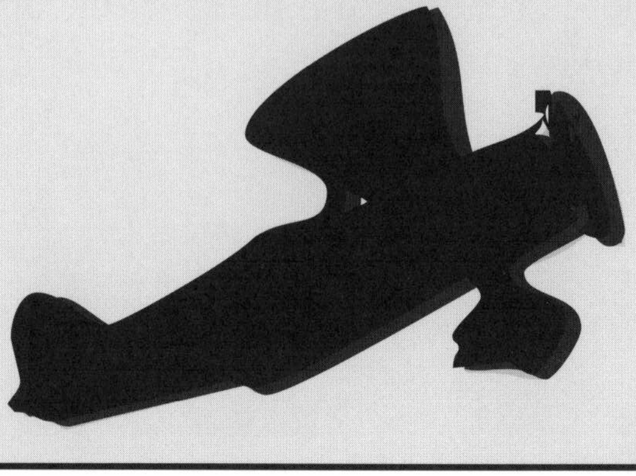

Good Luck and Godspeed! Polite applause follows...

EXPLANATION 4.6 Rubrics for Scoring Student Work

Rubrics contain explicit descriptions of what you expect to find in student work. You can use these scoring instruments to communicate your intentions to parents and students, and as tools for accurately gauging student performance. Ideally, students use the information in rubrics to set personal performance goals and to monitor their progress toward reaching these ends.

> *"The best rubrics are worded in a way that covers the essence of what we, as teachers, look for when we're judging quality..."*
>
> **Arter & McTighe, 2000**

The Internet is a treasure trove of information on the subject of rubrics. A quick search will yield hundreds of resources, some of which can be used immediately with your classes. In this Extension, you will access the Internet to gain a better understanding of rubrics and apply this knowledge to design and evaluate rubrics.

DIRECTIONS

PART A: HOLISTIC AND ANALYTIC RUBRICS

1. Study these rubrics carefully on the following page:

   - Figure 4.11 Holistic Rubric
   - Figure 4.12 Analytic Rubric

Rubrics like these must look familiar to you because such scoring guides have become extremely popular. The example in Figure 4.11 is called a holistic rubric; Figure 4.12 illustrates an analytic rubric.

2. Use the Compare and Contrast Thinking Diagram (Figure 4.13) to analyze the differences and similarities between the two rubrics. The way to use this diagram most effectively is to identify the essential attributes that distinguish the two types of rubrics. Remember, the purpose of this work is not to evaluate rubrics, but to differentiate among their features.

3. After you complete the Compare and Contrast Thinking Diagram, write a brief paragraph that summarizes your findings.

FIGURE 4.11 HOLISTIC SCORING RUBRIC

CIRCUITS AND PATHWAYS **PERFORMANCE ASSESSMENT**	
Outstanding **4**	All questions are answered completely and acceptably; gives logical explanations on questions 8, 10, and 11; draws a complete diagram of a circuit; constructs a table that clearly distinguishes conductors from nonconductors; shows understanding of electricity and conductivity.
Competent **3**	Most questions are answered completely and acceptably; explanations on questions 8, 10 and 11 may be unclear; draws an acceptable current; table may be less sophisticated (i.e., list format); shows understanding of electricity and conductivity.
Satisfactory **2**	Many incomplete or unacceptable responses; explanations on questions 8, 10, and 11 are very limited; diagram of circuit may be incomplete or missing; shows some understanding of electricity but not conductivity; attempts to answer questions on both pages.
Serious Flaws **1**	Response severely limited (only diagram or few answers); incomplete answers with no explanation or rationale; no evidence of understanding of the concept of electricity or conductivity; at least one correct answer other than the diagram.
No Attempt **0**	

Source: Brown, J. H. & Shavelson, R. J. (1996). *Assessing hands-on science*. Thousand Oaks, CA: Corwin.

FIGURE 4.12 ANALYTIC SCORING RUBRIC

MATHEMATICS EXTENDED-RESPONSE ITEMS

	MATHEMATICAL KNOWLEDGE Knowledge of mathematical principles and concepts that result in a correct solution to a problem.	STRATEGIC KNOWLEDGE Identification of important elements of the problem and the use of models, diagrams, symbols, and/or algorithms to systematically represent and integrate concepts.	EXPLANATION Written explanation and rationales that translate into words the steps of the solution process and provide justification for each step. Though important, length of the response, grammar, and syntax are not the critical elements of this dimension.
4	• Shows complete understanding of the problem's mathematical concepts and principles • Uses appropriate mathematical terminology and notations including labeling answer, if appropriate • Executes algorithms completely and correctly	• Identifies all the important elements of the problem and shows complete understanding of the relationships among elements • Reflects an appropriate and systematic strategy for solving the problem • Solution process is nearly complete	• Gives a complete written explanation of the solution process employed; explanation addresses both *what* was done and *why* it was done • If a diagram is appropriate, there is a complete explanation of all the elements in the diagram
3	• Shows nearly complete understanding of the problem's mathematical concepts and principles • Uses nearly correct mathematical terminology and notations • Executes algorithms completely; computations are generally correct but may contain minor errors	• Identifies most of the important elements of the problem and shows general understanding of the relationships among them • Reflects an appropriate strategy for solving the problem • Solution process is nearly complete	• Gives a nearly complete written explanation of the solution process employed, or explains both *what* was done and begins to address *why* it was done • May include a diagram with most of the elements explained
2	• Shows some understanding of the problem's mathematical concepts and principles • May contain major computational errors	• Identifies some important elements of the problem but shows only limited understanding of the relationship among them • Appears to reflect an appropriate strategy but the application of strategy is unclear, or a related strategy is supplied logically and consistently • Gives some evidence of a solution process	• Gives a written explanation of the solution process employed; either explains *what* was done or addresses *why* it was done; explanation is vague or difficult to interpret • May include a diagram with some of the elements explained
1	• Shows limited to no understanding of the problem's mathematical concepts and principles • May misuse or fail to use mathematical terms • May contain major computational errors	• Fails to identify important elements or places too much emphasis on unimportant elements • May reflect an inappropriate or inconsistent strategy for solving the problem • Gives minimal evidence of a solution process; process may be difficult to identify	• Gives minimal written explanation of the solution process; may fail to explain *what* was done or *why* it was done • Explanation does not match presented solution process • May include minimal discussion of elements in diagram; explanation of significant elements is unclear
0	• No answer attempted	• No apparent strategy	• No written explanation of the solution process is provided

Source: From *Illinois Standards Achievement Test: Sample Mathematics Materials 2000* (p. 59). © 2000, Illinois State Board of Education. Used with permission.

FIGURE 4.13 COMPARE/CONTRAST THINKING DIAGRAM: RUBRICS

Holistic		Analytic

HOW ALIKE?

-
-
-

HOW DIFFERENT?

Holistic	WITH REGARD TO ATTRIBUTE	Analytic
	• ⟷	
	• ⟷	
	• ⟷	

Adapted from Parks, S. & Black, H. (1992). *Organizing thinking*. Pacific Grove, CA: Critical Thinking Press and Software.

EXPLANATION 4.6 Rubrics for Scoring Student Work

DIRECTIONS

PART B: DISCOVERY GUIDE: RUBRIC STRUCTURE

1. For this section, you should refer to the analytic rubric shown in Figure 4.12. By carefully scrutinizing this rubric, you should be able to identify and investigate its three basic components.

2. Look at the left-hand column. These items are referred to as *performance levels*. There is a growing tendency to refer to these levels as *variations*.

- What is one of the performance levels for this rubric?
- How are performance levels for a rubric derived?
- What information do performance levels in a rubric convey to students?
- How do performance levels in a rubric assist the teacher?

3. Examine the top row of the rubric. These are called *performance criteria* or *descriptors*.

- What is one of the performance criteria for this rubric?
- How are performance criteria for a rubric derived?
- What information do performance criteria in a rubric convey to students?
- How do performance criteria in a rubric assist the teacher?

4. Examine the contents of the boxes in the rubric. These are the *performance indicators*.

- What is one of the performance indicators for this rubric?
- How are performance indicators for a rubric derived?
- What information do performance criteria in a rubric convey to students?
- How do performance criteria in a rubric assist the teacher?

EXPLANATION 4.6 Rubrics for Scoring Student Work

PART C: BUILDING YOUR RUBRIC

A growing number of software products, such as Rubric Maker™ and Rubricator™, are available commercially. These tools enable teachers to build their own rubrics quickly and easily. At the following website, you can access a free rubric generator: **teachers.teachnology.com/web_tools/rubrics**. Use the Scoring Guide for a Rubric (Figure 4.14) to evaluate a rubric you create on the site. You can use this freeware to construct your rubric, or, if you prefer, use the traditional paper-and-pencil method.

DIRECTIONS

Here is your assignment. In Part B of Explanation 4.5, you designed a performance task that was aligned with the learning goal that you have been following since Explanation 3.3. You also specified the criteria for success that you expect to find in student products or actions. These criteria provide a framework for constructing the rubric that accompanies your performance task. These expectations constitute the assessment's most important features and enable you to judge the quality of a piece of student work.

Here are some practical hints to remember as you create your rubric:

- Give the rubric a clear title.
- Choose performance levels that are consistent with your grading process. Make sure that these enable you to make clear distinctions among levels of quality.
- The highest ranking scoring levels typically represent the most desirable performance.
- Read the performance task carefully. What are the criteria for success? Use these as your rubric's descriptors.
- Performance indicators are the most important part of a rubric, but are the most difficult to articulate.
- Because the highest level is generally the easiest variation, you will probably enjoy quicker success if you begin here.
- Adjust the language among the variations so that distinctions among performance indicators are clear and unambiguous.

This final action completes your performance assessment. Remember, first drafts are just that, preliminary attempts! The quality of your rubrics will improve with experience and as you gain confidence. Before completing the Guide, you will have another opportunity to polish your draft of this performance assessment. You are now ready to proceed to the design of your standards-based instructional material.

REFLECTION 4.6 Rubrics for Scoring Student Work

Use the Scoring Guide for Evaluating a Rubric (Figure 4.14) to determine if your design is consistent with the framework described in the previous section. This figure incorporates most of the features associated with rubrics that effectively score student work.

1. Overall, how would you describe the quality of your rubric?

2. How would you describe the process of designing a rubric that works well with your students?

3. What are the major strengths of your rubric?

4. What are the areas that need improvement?

FIGURE 4.14 SCORING GUIDE FOR EVALUATING A RUBRIC

	Exemplary	Proficient	Satisfactory	Inadequate
Performance Levels (Variations)	• Levels precisely distinguish among possibilities.	• Levels clearly distinguish among possibilities.	• Levels hardly distinguish among possibilities.	• Levels fail to distinguish among possibilities.
Performance Criteria (Descriptors)	• Identify all major performance categories. • Distinctions among categories are clear and concise.	• Identify most major performance categories. • Distinctions among categories are generally clear.	• Identify some performance categories. • Distinctions among categories are imprecise.	• Many performance categories missing. • Distinctions among categories are confusing.
Performance Indicators	• Descriptors likely to be understood by all students. • Descriptors clearly connected to criteria. • Descriptors clearly differentiate among performance levels.	• Descriptors likely to be understood by most students. • Descriptors connected to criteria. • Descriptors distinguish among performance levels.	• Descriptors likely to be understood by some students. • Descriptors not clearly connected to criteria. • Descriptors do not generally distinguish among performance levels.	• Descriptors likely to be misunderstood by students. • Descriptors not clearly connected to criteria. • Descriptors fail to distinguish among performance levels.
Creativity	• Originality is high. • Likely to be attractive to most students.	• Somewhat original. • Likely to be attractive to many students.	• Originality is low. • Likely to be attractive to some students.	• Unoriginal. • Unlikely to be attractive to any student.

SUMMARY READING AND REFLECTION CHAPTER 4

Facing and Embracing the Assessment Challenge

Please answer these questions while you read the article, *Facing and Embracing the Assessment Challenge* (Damian, 2000). This paper can be found on the Internet at **www.enc.org**. Click on *encFocus*: Volume 7, number 2. You will find this article listed in the Table of Contents.

NOTE: If the paper is not available at the above web site, the reference is: Damian, C. (2000). Facing and embracing the assessment challenge. *encFocus,* 7, 2.

"Not everything that counts can be counted, and not everything that can be counted counts."

Albert Einstein

1. What do you think constitutes the major assessment challenge?

2. What is your understanding of the expression *embedded assessment*?

3. In today's classrooms, there is a trend toward student self-assessment. What are your thoughts on the following quote? "Effective learners operate best when they have insight into their own strengths and weaknesses ..." (Brown, 1994, p. 9).

SUMMARY READING AND REFLECTION CHAPTER 4

Facing and Embracing the Assessment Challenge

4. What is your reaction to the second Stiggins quote in the article?

5. What is the author's position on standardized testing? Does Danien's view complement or conflict with your own?

6. What is the most important statement in this article? Why did you choose this particular quote?

"The standards-based classroom is alive with student activity, is probably noisy, and may initially seem confusing to an outsider. But to those inside the classroom, this is a place where ideas are born, develop, and thrive."

Leonard, Penick & Douglas, 2002

SUMMARY READING AND REFLECTION CHAPTER 4

The Authentic Standards Movement and Its Evil Twin

No discussion about assessment would be complete without a reference to standardized testing and the issues that surround this controversial topic. Complete this reading response while you read The Authentic Standards Movement and Its Evil Twin. This article is found on the Internet at **www.pdkintl.org/kappan/ktho0101.htm**.

NOTE: If the paper is not available at the above web site, the reference is: Thompson, S. (2001). The authentic standards movement and its evil twin. *Phi Delta Kappan*, January, 358-362.

> *"Standardized testing has swelled and mutated, like a creature in one of those old horror movies, to the point that it now threatens to swallow our schools whole."*
>
> **Kohn, 2000**

1. Although there is a clear distinction between standards-based reform and standardized testing, people often use the terms interchangeably. What do you think accounts for this confusion?

2. What is the difference between a standards-based assessment and a standardized test?

3. Is teaching to the test the same as teaching to the standard? Why or why not?

SUMMARY READING AND REFLECTION

CHAPTER 4

The Authentic Standards Movement and Its Evil Twin

4. What do you see as your personal responsibility to students in terms of how well they perform on standardized tests?

5. Might the present emphasis on standardized testing jeopardize the standards reform movement? Explain your thoughts.

6. How do you intend to personally cope with the tension between standards for students and standardized testing?

GUIDED REFLECTION
CHAPTER 4

We believe that our fictional character Julian, introduced at the beginning of the chapter, represents a typical user of the Guide. Completing the activities in the Guide gave him a solid understanding of the standards reform movement and a strategy for using the national resource documents to clarify the intent of standards. Armed with this background, he immersed himself in a careful study of contemporary approaches to assessment. Both you and Julian are well prepared to apply your knowledge of national standards and assessment to implement changes in the classroom.

The Guide's format provides versatility. Figure 4.15 illustrates the three Extension options. If your interest is in developing your own standards-based curriculum materials, you should go directly to Chapter 5. If you need a systematic method to analyze and evaluate curricula, then Chapter 6 will apply. If you want a guide for mapping a comprehensive K-12 curriculum, then you should move on to Chapter 7.

Although we urge you to complete all activities in Chapters 1-4, specially tailored Learning Paths are presented in Chapters 6 and 7 for those who wish to restrict their use of the Guide's principles to curriculum evaluation or mapping. These streamlined applications of the Guide should provide sufficient preparation for these purposes.

FIGURE 4.15 EXTENSION OPTIONS

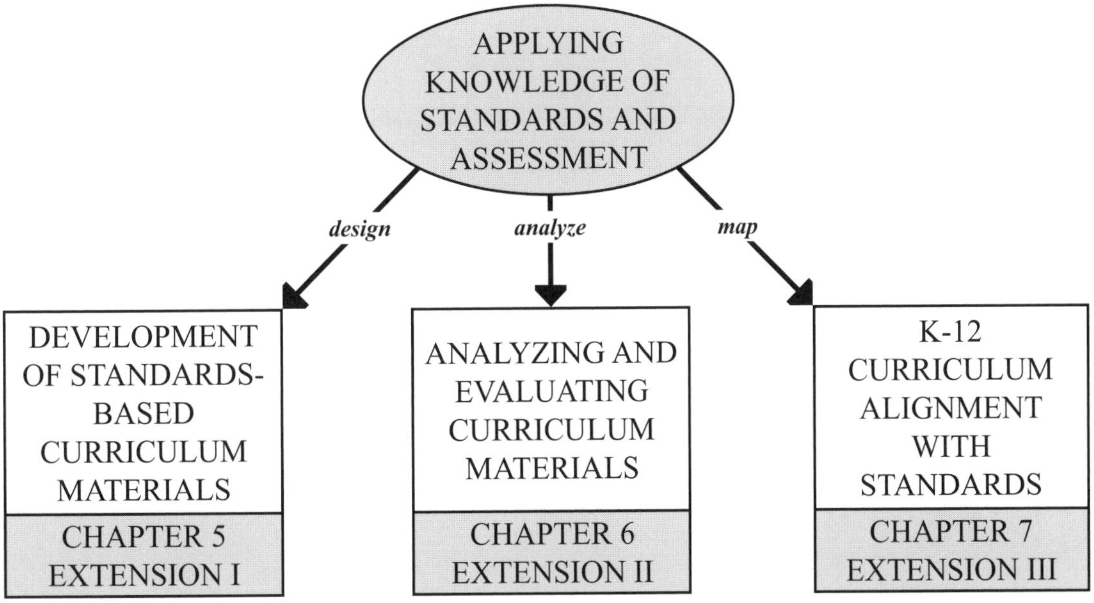

EXTENSION I: Developing Standards-Based Curriculum Materials

CHAPTER 5

> *"Begin with the end in mind."*
>
> **Covey, 1989**

In a learning cycle, Extensions provide learners with the opportunity to apply their new understanding in novel situations. As you will recall, standards present a vision for the *ends* of education; the resulting curriculum that teachers design and the instruction they provide are the *means* for achieving these desired results for students. And so, the ultimate yardstick for measuring if students are being provided with sufficient opportunities to meet standards is reflected by the curriculum and through instruction.

Chapter 5 will introduce and apply an instructional planning model that utilizes standards as the basis for deciding what students need to know or be able to do, assessing student performance, and making choices about learning experiences that students should be provided. We hope you will find that the significant difference between the standards-based procedures described in the Guide and typical lesson planning is the way that instruction and assessment are designed, organized, and sequenced. You will also use a procedure for deciding what lesson accommodations are required to make a curriculum accessible for all learners. Although you may find that completing this section requires considerable time and energy, you are well prepared for the task.

	LEARNING GOALS	**ACTIVITIES**
EXTENSION I	Identify the essential features of standards-based instruction.	5.1: Refining Your Performance Assessment
	Finalize the performance assessment.	
	Make necessary lesson accommodations for all learners.	5.2: Ensuring That Learning Experiences Are Accessible to All Students
	Select learning experiences that target a specific learning goal.	5.3: Designing Instruction: Standards-Based Templates
	Prepare standards-based instructional materials using a learning cycle.	

SUMMARY REFLECTION	*Self-Assessment of Standards-Based Materials*

EXTENSION 5.1 Refining Your Performance Assessment

Before selecting learning activities that your students will experience, you should complete one final review of the performance assessment designed in Chapter 4. Figure 5.1 provides a quick, simple, and effective procedure for critiquing your assessment instrument and for making required adjustments. Remember, your performance assessment is likely to remain in draft form until your curriculum material is finalized. After you implement the assessment, you will probably want to rewrite and further refine it after receiving students' feedback and examining their work.

> *"The way a lesson is organized provides a shell, or context, within which the teacher engages students in learning the subject."*
>
> **Stigler & Hiebert, 1999**

DIRECTIONS

Take a careful look at each of the following attributes of a well-designed performance assessment to rate your own assessment instrument. A rating of 5 is very good; 1 is poor. In this 5-point evaluation system, any feature that scores 4 or 5 is satisfactory in its current form. Revisions are advisable for any statement that receives a rating below 4.

To guide your analysis, ask yourself, "What features of this performance assessment should I alter to improve the quality and relevance to my students?" Your response to this question will dictate the specific changes that you make to the assessment.

FIGURE 5.1 ATTRIBUTES OF A WELL-DESIGNED PERFORMANCE ASSESSMENT

Alignment…	
Includes a careful correspondence between the assessment and the targeted learning goal(s).	⑤④③②①
Is free of bias.	⑤④③②①
Generates accurate and valid evidence that students have reached the standards.	⑤④③②①
Provides feedback for teacher to use in adjusting instruction.	⑤④③②①
The Task…	
Is clear to students.	⑤④③②①
Is appropriate for the grade level.	⑤④③②①
Is engaging.	⑤④③②①
Creates an authentic and relevant context.	⑤④③②①
Challenges student to apply new knowledge.	⑤④③②①
Is feasible to implement for teachers and students.	⑤④③②①
Makes accommodations for children with special needs.	⑤④③②①
The Criteria for Success…	
Describe specific criteria for the product or action.	⑤④③②①
Are clear about expectations for students.	⑤④③②①
The Scoring Guide…	
Has scoring variations that differentiate among performance levels.	⑤④③②①
Has performance criteria that are closely linked to the performance task.	⑤④③②①
Has clear indicators to distinguish different student performance levels.	⑤④③②①
Can be used for student self-assessment.	⑤④③②①

EXTENSION 5.2 Ensuring That Learning Experiences Are Accessible to All Students

Before implementing a lesson, teachers should conduct a final analysis to determine if any portion of this activity would be difficult or unsuitable for students with disabilities. The Individuals with Disabilities Education Act (IDEA) is a historic piece of special education legislation that ensures educational equity in the classroom. This requirement for equal access for all students, regardless of the nature of their disability, imposes special challenges in active learning environments such as those advocated by the Guide.

> *"Inclusive education means that all students regardless of their strengths or weaknesses become part of the school community."*
>
> **Renaissance Group, 1999**

When teachers systematically adapt their instruction to make educational opportunities available to everyone, the process is called *accommodation*. For example, providing large-print materials can accommodate a student who has difficulty reading standard-sized text. The Student Activities Profile Model (DO-IT, 2000) illustrated in Figure 5.2 is essentially a task analysis procedure. Each lesson is evaluated in terms of the unique set of cognitive, sensory, and physical skills required for its successful completion. This analytic framework identifies the necessary accommodations for making these learning experiences accessible to all learners. You should review each of the lessons in your curriculum package using the chart below. Then, when students complete the activities, the accommodations you have made ensure that this learning experience is within reach of everyone.

FIGURE 5.2 STUDENT ACTIVITIES PROFILE MODEL

LESSON			
DESCRIPTION OF TASKS			
Required Physical Skills			
Required Sensory Skills			
Required Cognitive Skills			
Features of Task That Require Accommodation			
Accommodation Options			

EXTENSION 5.3 Designing Instruction: Standards-Based Templates

> *"Imagine the lesson as a story. Well-formed stories consist of a sequence of events that fit together to reach the final conclusion. Ill-formed stories are scattered sets of events that don't seem to connect."*
>
> **Stigler & Hiebert, 2000**

Think of the number of lessons and units that you have prepared during your teaching career. Your experience is immense. In this section, you have the opportunity to apply this expertise.

The framework of your standards-based material was drawn from the curriculum congruence model described in Explanation 4.1. The process for selecting and organizing the learning activities in your standards-based material originates in two activities from the Guide: the Content Clarification in Explanation 3.3, and the performance assessment you designed for Explanations 4.5 and 4.6. Because little here is new in terms of effective teaching, the following planning guidelines should look familiar. Regardless of the stage in instructional design, the focus remains the same, namely, improving student performance.

Because this section integrates everything that you have previously experienced in the Guide, the instructions are minimal. However, even the most experienced teachers should follow the recommended and logical stepwise arrangement. You will discover that the procedure often loops back and that the steps are actually overlapping. As your level of confidence with standards-based instructional design rises, you will undoubtedly find ways to streamline the process and use it more efficiently in your everyday routine. The following descriptions explain the learning cycle, provide helpful design tips, and offer suggestions for using each of the templates.

FIGURE 5.3: LEARNING CYCLE MODEL OF INSTRUCTION

The learning cycle provided a framework for the Guide. Hopefully, you concluded that a learning cycle makes good sense as a strategy for organizing instruction that fosters conceptual change. We encourage you to select and sequence the learning activities in your materials on the basis of where they best fit within the learning cycle format. Figure 5.3 on page 92 illustrates the essential features of the learning cycle and provides guiding questions to determine the most appropriate stage for including an activity. Each stage is referenced to pages in the Guide where you can review details about the individual components of this model.

FIGURE 5.4: SUGGESTIONS FOR DESIGNING YOUR LESSON

These recommendations come from teachers who have applied the Guide's principles to develop standards-based materials for their classrooms. These hints can simplify and streamline your planning (see page 93).

EXTENSION 5.3 Designing Instruction: Standards-Based Templates

FIGURE 5.5: GRAPHIC ORGANIZER FOR A STANDARDS-BASED UNIT

This diagram illustrates the overall process for preparing instructional materials that originate in standards and learning goals. Because of your experience with the Guide, this planning approach should look very familiar. Select a learning goal and then prepare an aligned assessment. Finally, design an instructional scope and sequence that enables students to meet the targeted learning goal. Apply a learning cycle to organize the learning experiences.

FIGURE 5.6: STANDARDS-BASED UNIT: STANDARDS AND PERFORMANCE ASSESSMENT

The curriculum materials that you are developing are anchored in a specific standard and learning goal. You investigated the intent of this learning goal in Explanation 3.3. Summarize the information from your Content Clarification in Part 2 of Figure 5.6. In Extensions 4.4 and 4.5, you prepared an assessment that was aligned with the learning goal that you selected. Summarize the elements of your performance assessment in Part 2 of Figure 5.6.

FIGURE 5.7: PART A—STANDARDS-BASED LESSON DESIGN: FOR THE TEACHER

Figure 5.7 on page 96 is the principal planning tool for your individual lessons. The Guide has previously addressed each item of this template in detail. You should complete one of these templates for each lesson in your unit. Obviously, not all sections of this template will be appropriate for every lesson.

FIGURE 5.8: PART B—STANDARDS-BASED LESSON DESIGN: FOR THE STUDENT

In a standards-based approach to teaching and learning, it is generally advisable to make students aware of the targeted learning goals for most lessons. Depending on the nature of a particular lesson, it may not be fitting to complete certain sections of this template (page 97).

FIGURE 5.9: COVER SHEET FOR A STANDARDS-BASED UNIT

This template (page 98) is equivalent to an abstract or overview of your unit. For this reason, it should be the final piece that you complete. Any teacher or administrator should be able to scan this and get a clear and complete picture of your instructional plan's scope and sequence.

FIGURE 5.3 LEARNING CYCLE MODEL OF INSTRUCTION

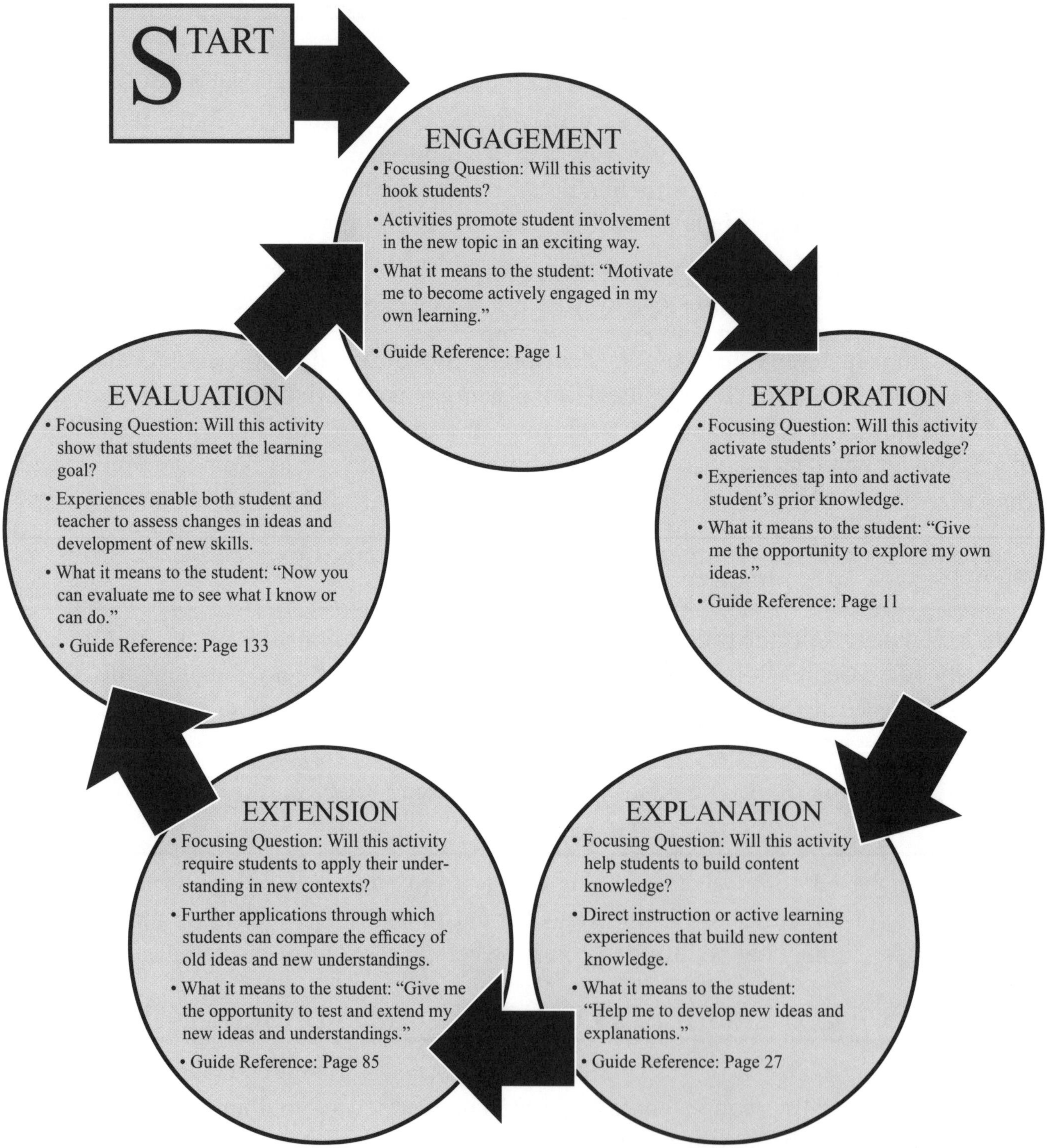

START
ENGAGEMENT
• Focusing Question: Will this activity hook students?
• Activities promote student involvement in the new topic in an exciting way.
• What it means to the student: "Motivate me to become actively engaged in my own learning."
• Guide Reference: Page 1
EXPLORATION
• Focusing Question: Will this activity activate students' prior knowledge?
• Experiences tap into and activate student's prior knowledge.
• What it means to the student: "Give me the opportunity to explore my own ideas."
• Guide Reference: Page 11
EXPLANATION
• Focusing Question: Will this activity help students to build content knowledge?
• Direct instruction or active learning experiences that build new content knowledge.
• What it means to the student: "Help me to develop new ideas and explanations."
• Guide Reference: Page 27
EXTENSION
• Focusing Question: Will this activity require students to apply their understanding in new contexts?
• Further applications through which students can compare the efficacy of old ideas and new understandings.
• What it means to the student: "Give me the opportunity to test and extend my new ideas and understandings."
• Guide Reference: Page 85
EVALUATION
• Focusing Question: Will this activity show that students meet the learning goal?
• Experiences enable both student and teacher to assess changes in ideas and development of new skills.
• What it means to the student: "Now you can evaluate me to see what I know or can do."
• Guide Reference: Page 133

FIGURE 5.4 SUGGESTIONS FOR DESIGNING STANDARDS-BASED MATERIALS

- Build your materials around the learning cycle model (see Figure 5.3).
- Complete at least one lesson template for each stage in the learning cycle.
- Don't start from scratch. Begin by reviewing the learning activities that you currently use to teach this topic.
- Continuously reference the targeted learning goals and the performance assessment criteria to evaluate your selection of learning activities.
- Apply the Focusing Questions in the Learning Cycle model to determine the most appropriate application of an activity.
- Always ask, "Does this activity really contribute to the student's ability to achieve the learning goals on which the assessment is based?"
- If an activity is a good one, but doesn't really address the targeted learning goals, then consider using it only for enrichment purposes.
- Don't hesitate to discard an activity that you've been using for years when it doesn't contribute to achieving the targeted learning goals. Reeves (1998, p. 49) refers to this as "pulling the weeds."
- Design the scope and sequence of your lessons after you've identified possible learning activities.
- After you've developed the sequence of lessons, ask, "Does this set of learning activities fully complement the performance assessment?"
- Congratulate yourself on a job well done!

FIGURE 5.5 GRAPHIC ORGANIZER FOR A STANDARDS-BASED UNIT

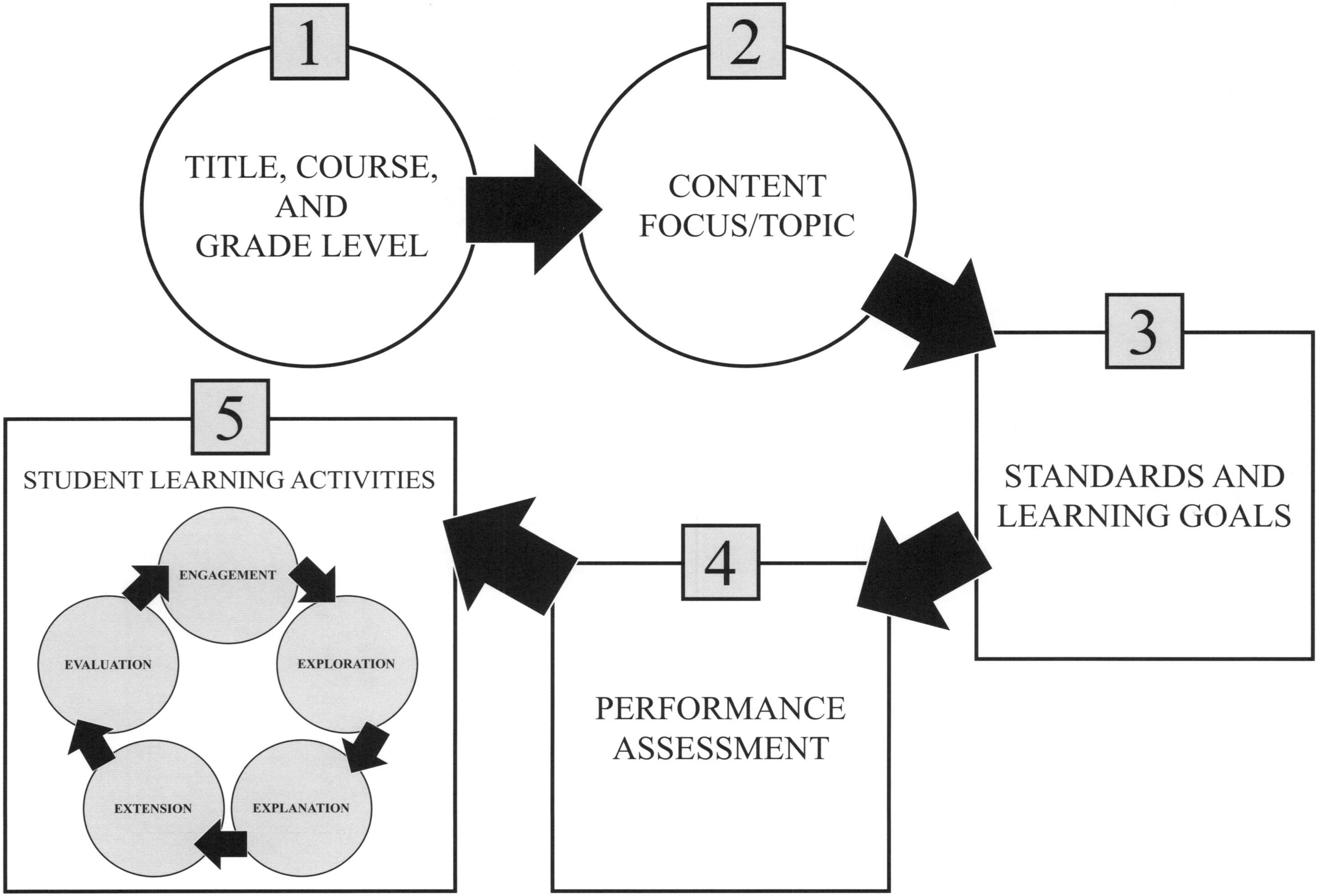

FIGURE 5.6 STANDARDS-BASED UNIT: STANDARDS AND PERFORMANCE ASSESSMENT

Unit Title	
Content Focus/Topic	
Grade Level/Course	
Standard(s)	
Learning Goal(s)	

1 Review of Standards and Learning Goals

Suggestion: refer to Figure 3.5: Content Clarification of a Learning Goal

Content	What exactly should the student know after completing this unit of instruction?	Skills	What exactly should the student be able to do after completing this unit of instruction?

2 Review of Performance Assessment

Suggestion: refer to Explanations 4.5: Building Your Performance Assessment and 4.6, Part 3: Building Your Rubric

Task Description	What is the role/persona of the student or group, the target behavior, the audience, and the product or action?
Criteria for Success	What are the expected characteristics of the product or action?
Scoring Guide	What specific indicators of student performance are evaluated?

FIGURE 5.7 PART A—STANDARDS-BASED LESSON DESIGN: FOR THE TEACHER

Learning Stage Cycle

Check the stage that best applies to this lesson.

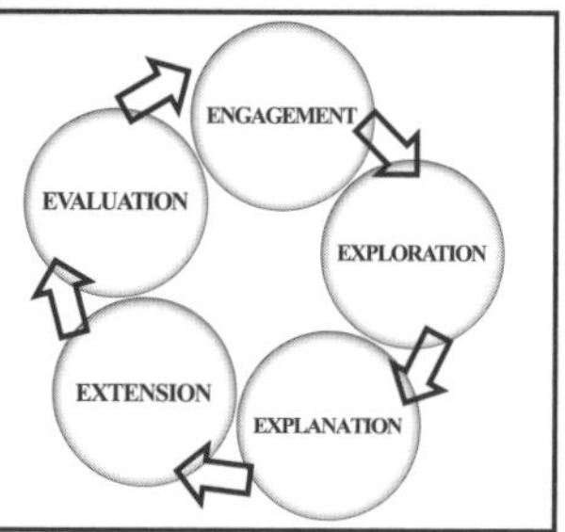

1 Lesson Title

2 Lesson Overview

3 Instructional Goals

•	•
•	•

4 Formative Assessment(s)

5 Lesson Organization

Opening the Lesson	Developing the Lesson	After the Lesson
Time Allocation:	Time Allocation:	Time Allocation:

6 Instructional Strategies

-
-
-

7 Teaching Resources

-
-
-

8 Potential Accommodations for Special Learners

Type of Disability	Accommodation
•	•
•	•

9 Enrichment Activities

-
-
-

FIGURE 5.7 PART B—STANDARDS-BASED LESSON DESIGN: FOR THE STUDENT

1 Lesson Title

2 Targeted Learning Goals

-
-
-
-

3 Assessment

4 Description of Learning Activity

5 Materials and Equipment

FIGURE 5.8 COVER SHEET FOR A STANDARDS-BASED UNI

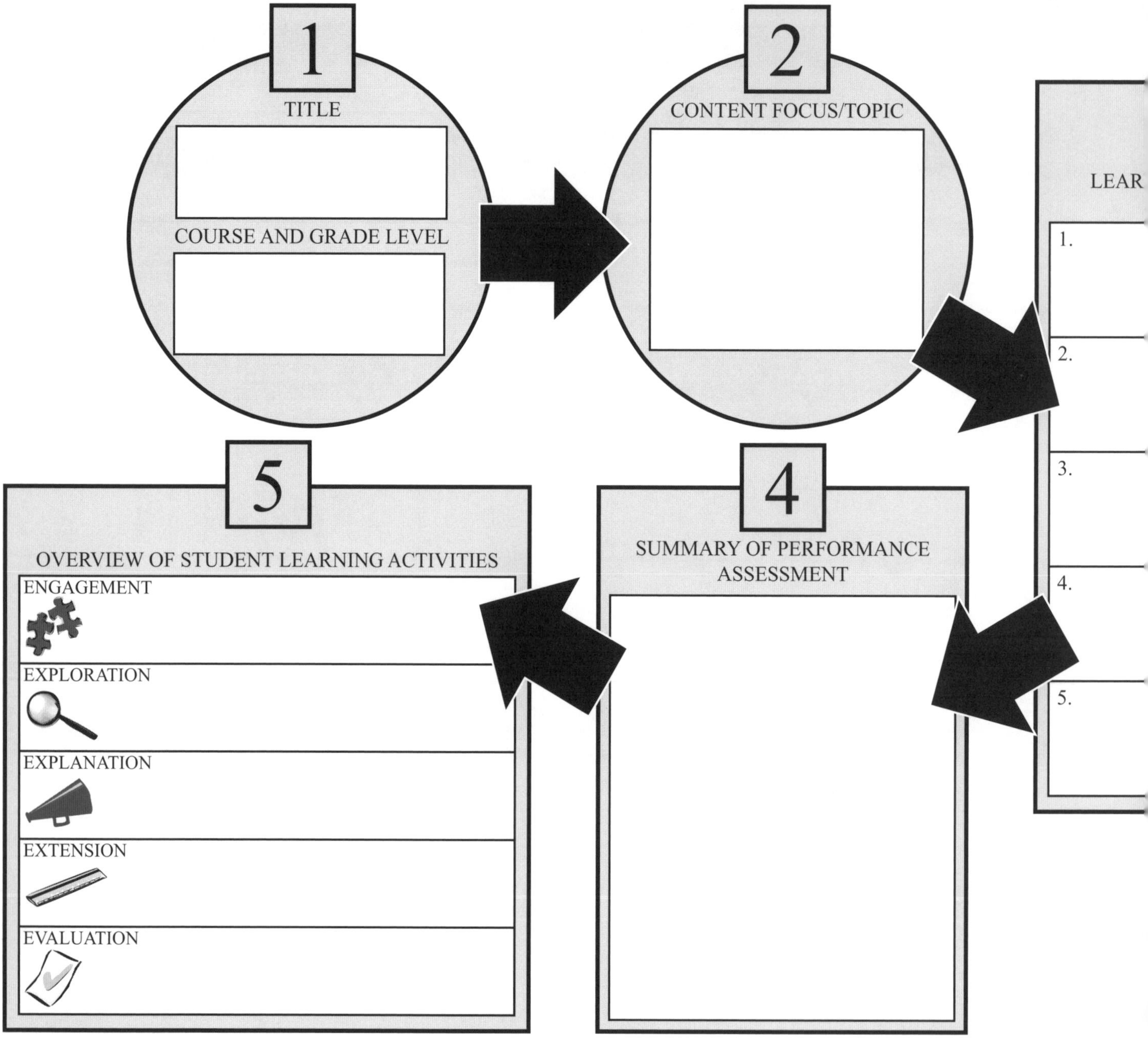

SUMMARY REFLECTION

CHAPTER 5

Use the rubric referenced in Figure 4.8 (page 66) as a self-assessment for your standards-based materials.

"While policy emphasizes standards and assessment, it is up to teachers to figure out how standards should be implemented in the classroom."

Madfes & Muench, 1999

1. How did you score your unit in terms of congruence?

2. How did you rate your performance assessment? What are its principal strengths? Weaknesses?

3. How did you rate your instructional plan? What are its principal strengths? Weaknesses?

4. What is your overall reaction to the backward design method used to prepare this standards-based material?

SUMMARY REFLECTION

CHAPTER 5

5. What do you think are the principal advantages of following a design path from standards to assessment to the design of instruction?

6. What do you think are the principal disadvantages of following a design path from standards to assessment to the design of instruction?

7. In terms of designing curriculum and instruction, what does it really mean to begin with the end in mind?

8. Do you think that it is realistic to expect that individual teachers will develop a large collection of their own standards-based curricula? Explain your answer.

EXTENSION II: Analyzing and Evaluating Curriculum Materials

CHAPTER 6

> *"The reality is that teaching is a full-time job, and it is very unlikely that full-time teachers can also develop all the high-quality curriculum materials they will need, any more than full-time practicing physicians are likely to come up with all the procedures, equipment, and medications they need."*
>
> **Tucker & Codding, 1998**

Since the advent of standards, an abundance of curriculum materials claiming to be "standards-based" have appeared. Project 2061 conducted an extensive analysis of algebra, biology, and middle school science textbooks in terms of their specific content goals and suggested approaches to instruction (Kesidou, 2001). Their research revealed that most textbooks were woefully mismatched with the current recommendations associated with standards-based teaching and learning.

In the introduction to the Guide, we stated our view that "learning about standards is neither simple nor direct." The same argument could be made for problems associated with evaluating and selecting standards-based curricula for your course or grade level. Again, professional development may offer the critical component for gaining the necessary skills for identifying materials that complement standards-based instruction.

	LEARNING GOALS	ACTIVITIES
EXTENSION II	Recognize how ideas about national standards, constructivist teaching and learning, performance assessment, and curriculum design provide a framework for analyzing instructional materials.	6.1: Preparing to Analyze Curriculum Materials
	Apply the Guide's major principles to analyze curriculum materials.	6.2: Curriculum Analysis and Evaluation
	Utilize the Curriculum Analysis Rubric.	
	Interpret data from the Curriculum Analysis Rubric to evaluate curriculum materials.	

SUMMARY REFLECTION	*Standards-Based Curriculum Evaluation*

GUIDED REFLECTION

CHAPTER 6

To illustrate the textbook adoption issues that schools are facing, consider the following scenario. Maria has just been appointed as the new Director of Curriculum and Instruction for her school district. Her first major leadership responsibility is to prepare teachers for making informed decisions about selecting new programs and materials. She wants teachers to be guided by ideas that are consistent with standards-based approaches to K-12 mathematics and science education.

From her previous experience with the Guide, she knows that the tools in this book can help teachers understand the principles of standards-based reform and discover how to apply constructivist teaching and learning practices in their classrooms. The Guide includes a section called Analyzing Curriculum Materials that seems well suited to the needs of her Textbook Selection Committee. When planning the workshop for her group, Maria wondered, "Which activities from the Guide are essential for gaining the prerequisite background to analyze textbook materials from a standards-based perspective?"

We applied the same logic that this fictitious Curriculum Director would use in developing a framework for Extension II. If you are like most teachers, you will never be fully responsible for developing an entire standards-based curriculum for your course or grade level. However, it is reasonable to expect that you be prepared to recognize and evaluate the features of classroom materials from a standards-based perspective. This extension provides tools for accomplishing this task.

It is important to note that your school district or state may already have an established textbook review process. The curriculum analysis approach presented here is based solely on the principles of standards-based reform embedded throughout the Guide. The recommended procedure may omit factors, such as content accuracy, that your district or state deems important to evaluate. Potential dilemmas can be resolved by applying one of two alternatives. Use the Guide's suggested approach as an initial screening tool to check for congruence with standards, or modify the Curriculum Analysis Rubric to accommodate your additional evaluative criteria. The References and Further Readings section lists several other curriculum analysis methods that may be more consistent with your particular needs than the procedure presented in the Guide.

EXTENSION 6.1 Preparing to Analyze Curriculum Materials

> *"Conditions surrounding materials selection may lead evaluators to review materials superficially and choose those that look attractive, appear to reduce budget outlays, or simplify teachers' roles. For this reason, building a local capacity to select instructional materials that support the goals of state and national standards is of paramount importance."*
>
> **National Research Council, 1999**

Individual teachers generally find it impractical to apply procedures such as those introduced in Extension I for developing standards-based curriculum materials in the sense of a complete and comprehensive course of study. Most teachers lack the time, inclination, or expertise needed for conducting such massive undertakings. Instead, they rely upon commercially available curriculum materials. But teachers need to be critical consumers of educational materials. Becoming an informed evaluator of standards-based curricula takes practice. It happens best through rigorous study aimed at developing an in-depth and thorough understanding of the principles associated with curriculum congruence, constructivist teaching and learning, and assessment. The seven steps in the Learning Path outlined below provide a framework drawn from the Guide that can prepare a team of teachers to analyze and evaluate instructional materials for their particular courses or grade levels.

NOTE: For the purposes of the Guide, the expression "instructional material" refers to anything that a teacher might use to assist with teaching science or mathematics. Examples of materials include a textbook, web site, laboratory manual, instructional kit, etc.

DIRECTIONS

If you completed the Guide in its entirety, then a cursory review of the sections referenced in this Learning Path should suffice for conducting the curriculum analysis. If not, you should fully complete each of the recommended sections in this Learning Path before beginning your curriculum materials analysis and evaluation. The Curriculum Analysis Focusing Questions provide connections between the particular steps in the Learning Path and the rubric that you will use for the analysis in Extension 6.2. For each step, you should prepare a statement that summarizes the Central Idea of the activity. This will help you to ascertain that the materials address the *intent* of the standard.

EXTENSION 6.1 Discovery Guide: Curriculum Materials Analysis Learning Path

STEP ONE

ENGAGEMENT 1.2	Beliefs and Attitudes About Constructivism and Standards-Based Learning

Curriculum Analysis Focusing Question for Engagement 1.2: Which characteristics of constructivist teaching and learning are evident in these curriculum materials?

Central Idea of Engagement 1.2:

> *"Teachers work with teaching tools. They are unlikely to change their practice significantly in the absence of supporting curricular materials."*
>
> **Bransford, Brown, & Cocking, 1999**

STEP TWO

EXPLORATION 2.3	Getting to Know the National Standards Resources

Curriculum Analysis Focusing Question for Exploration 2.3:

What information do the national standards resources provide about the learning goal that you are examining in these curriculum materials?

Central Idea of Exploration 2.3:

STEP THREE

EXPLANATION 3.1	Content Clarification of a National Standard

Curriculum Analysis Focusing Questions for Explanation 3.1:

- How consistent are these curriculum materials with recommendations found in the national standards?
- Do the materials incorporate the research findings associated with the content standard that you are examining?
- Is the intent of the standard fully addressed?

Central Idea of Explanation 3.1:

STEP FOUR

EXPLANATION 4.1	Congruence With Standards

Curriculum Analysis Focusing Question for Explanation 4.1:

What evidence do these curriculum materials contain for congruence among standards, assessment, and instruction?

Central Idea of Explanation 4.1:

EXTENSION 6.1 Discovery Guide: Curriculum Materials Analysis Learning Path

STEP FIVE

EXPLANATION 4.2	Traditional Versus Standards-Based Instruction	Central Idea of Explanation 4.2:
Curriculum Analysis Focusing Question for Explanation 4.2: What is the evidence that these curriculum materials incorporate the essential features of standards-based curriculum design?		

STEP SIX

EXPLANATION 4.3	The Language of Assessment	Central Idea of Explanation 4.3:
Curriculum Analysis Focusing Question for Explanation 4.3: Do these curriculum materials employ assessment practices that clearly demonstrate standards-based performances in the classroom?		

STEP SEVEN

EXTENSION 5.3	Designing Instruction: Standards-Based Templates	Central Idea of Extension 5.3:
Curriculum Analysis Focusing Question for Extension 5.3: Do the learning experiences and instructional approaches found in these curriculum materials effectively enable a student to meet the content standard that you are examining?		

REFLECTION 6.1 Preparing to Analyze Curriculum Materials

1. How important is it for persons who are participating in a textbook review and selection process to be knowledgeable in the principles of standards-based reform?

2. Why were the series of steps in this Learning Path considered to be essential for completing the analysis and evaluation of curriculum materials?

3. How did identifying the Central Ideas in this Learning Path provide you with background knowledge for the upcoming curriculum analysis and evaluation?

4. Which is the most important Curriculum Analysis Focusing Question? Explain your answer.

EXTENSION 6.2 Curriculum Analysis and Evaluation

> *"To achieve the learning goals of the Standards or Benchmarks, students and teachers must be provided with instructional materials that reflect these standards."*
>
> **National Research Council, 1999**

Earlier, you discovered that evaluation meant "the interpretation of assessment data, based on a set of criteria to judge." Evaluations are judgments that are based on data derived from an analysis of something. Three factors determine our ability to evaluate: the amount of available information, its quality, and the analytic tools and skills at our disposal. The process of generating and analyzing information provides a basis for making informed decisions about the items for your review.

Because it would be far too time-consuming for a selection committee to carefully scrutinize textbooks in their entirety, we recommend that as your first step, you select one or two learning goals within a particular standard and target them exclusively during your review. Use these same learning goals to analyze every material under consideration for your course or grade level. Our assumption is that all learning goals are of equal importance and that, within a textbook, all learning goals are addressed in a similar and regular fashion. This approach ensures that your group's cross-comparisons will be fair, focused, and consistent.

PART A: CURRICULUM ANALYSIS

For a textbook publisher to make the legitimate claim that a curriculum is standards-based, the materials must do more than simply *cover* particular topics referenced in the content standards that are found in the national documents. The Congruence Triangle (Explanation 4.1) illustrated that to achieve genuine alignment, the *intent* of the learning goals and the underlying ideas that these support must provide the principal basis for developing the curriculum. As you discovered earlier, the underlying meaning of a learning goal is best revealed through a Content Clarification (Explanation 3.3). This pre-planning strategy results in an in-depth understanding that serves as a foundation for your curriculum analysis and evaluation.

Extension 6.2 introduces the Curriculum Analysis Rubric (Figure 6.2 on page 110), a tool that is structured around the Focusing Questions found in Extension 6.1. This is why it was important for you to complete all seven steps in the Learning Path. These activities gave you a solid understanding of standards-based curriculum and instruction. Descriptors in the Curriculum Analysis Rubric were derived from the Central Ideas and principles of standards-based reform found in the Learning Path.

EXTENSION 6.2 Curriculum Analysis and Evaluation

Every member of the Curriculum Review Team should feel fully prepared to apply this rubric to gather data about any learning material in terms of its consistency with contemporary principles of educational reform.

DIRECTIONS

1. Before using the rubric, do the following:

- Carefully review the rubric to ensure that every member of the group understands its principal features.
- If necessary, modify the rubric by adding descriptors that make it consistent with your district or state's established curriculum review format.
- Identify the specific standard and learning goals that you have chosen to investigate in the collection of curriculum materials under review.
- Complete a Content Clarification for these learning goals as you did in Explanation 3.3.

2. As a team, you should review the curriculum materials carefully. Identify *every* location in the material that references the various descriptors found in the Curriculum Analysis Rubric associated with this learning goal. You will again use the term *sightings* for places in the text where connections with the learning goal exist. In Figure 6.2, you can list the sightings according to the descriptors found in the rubric. Remember, it will not be the *number* of sightings for a descriptor that determines its ultimate rating, but rather the nature of the content included in these sightings.

3. Read all of these sightings thoroughly by descriptor row before completing your analysis using the rubric.

4. Apply the Curriculum Analysis Rubric (Figure 6.1) to: (a) judge this curriculum material in terms of the learning goals on which you focused, and (b) describe its fidelity to the Central Ideas of standards-based reform that are underscored in the Guide.

5. Summarize your findings by completing the Evaluations in Parts B and C on page 111.

FIGURE 6.1 CURRICULUM ANALYSIS RUBRIC

INDICATORS	VARIATIONS			
	EXEMPLARY MATERIAL	MATERIAL MEETS THE STANDARD	MATERIAL APPROACHES STANDARD	MATERIAL BELOW ACCEPTABLE STANDARD
Standards (EXPLORATION 2.3, EXPLANATION 3.1)	• Standards and learning goals clearly identified. • Intent of standards and learning goals are fully addressed.	• Standards and learning goals mentioned. • Intent of standards and learning goals touched upon.	• Standards and learning goals poorly referenced. • Topic but not intent of standards and learning goals emphasized.	• Standards and learning goals not identified. • Topic is touched upon.
Assessment Plan (EXPLANATION 4.3)	• Clear and precise links to the standards and learning goals. • Clear and concise performance task. • Highly appropriate criteria for success. • Valid scoring system.	• Some links to the standards and learning goals. • Clear performance task. • Appropriate criteria for success. • Scoring system generally valid.	• Partial links to the standards and learning goals. • Unclear performance task. • Unclear criteria for success. • Scoring system somewhat valid.	• No links to the standards and learning goals. • No performance task. • No criteria for success. • No scoring system.
Instructional Plan (EXPLANATION 4.2, EXPLANATION 5.3)	• Always developmentally appropriate. • Uses a broad variety of instructional strategies. • Active and engaging learning activities. • Extremely well-organized scope and sequence. • Always emphasizes depth over breadth of content.	• Generally developmentally appropriate. • Uses a variety of instructional strategies. • Engaging learning activities. • Logical scope and sequence. • Sometimes emphasizes depth over breadth of content.	• Little attention to developmental concerns. • Little variety in instruction. • Mildly engaging learning activities. • Poorly organized scope and sequence. • Seldom emphasizes depth over breadth of content.	• No attention to developmental concerns. • No variety in instruction. • Unengaging learning activities. • Disorganized scope and sequence. • Content coverage emphasized.
Instructional Approaches (ENGAGEMENT 1.2)	• Consistently engages student's prior knowledge. • Formative assessments frequently included. • Provides a strongly scaffolded framework for conceptual development. • Always encourages reflective thinking.	• Sometimes engages student's prior knowledge. • Some formative assessments included. • Provides a basic framework for conceptual development. • Sometimes encourages reflective thinking.	• Seldom engages student's prior knowledge. • Few formative assessments included. • Provides a weak framework for conceptual development. • Seldom encourages reflective thinking.	• Never engages student's prior knowledge. • No formative assessments included. • No framework for conceptual development provided. • Reflective thinking not mentioned.
Curriculum Congruence (EXPLANATION 4.1)	• Unit is fully aligned with standards and learning goals. • Assessment is fully linked to learning goals. • Instruction is fully matched to standards, learning goals and assessment plan.	• Unit is generally aligned with standards and learning goals. • Assessment is generally linked to learning goals. • Instruction is generally matched to standards, learning goals and assessment plan.	• Unit is partially aligned with standards and learning goals. • Assessment is partially linked to learning goals. • Instruction is partially matched to standards and learning goals and assessment plan.	• Unit is not aligned with standards and learning goals. • Assessment is not linked to learning goals. • Instruction is not matched to standards, learning goals and assessment plan.

FIGURE 6.2 SIGHTINGS FOR RUBRIC DESCRIPTORS

Title of Curriculum Material						
Publisher						
Standard and Learning Goals Analyzed	• • •					
Connections With Standards and Learning Goals						
Assessment Strategies						
Characteristics of Instructional Plan						
References to Instructional Approaches						
Illustrations of Congruence						
Additional Descriptor						
Additional Descriptor						

EXTENSION 6.2 Curriculum Analysis and Evaluation

PART B: CURRICULUM EVALUATION

Title of Curriculum Material			
Publisher		Edition	

How did you rate the overall quality of this curriculum material in terms of its consistency with standards-based principles?

	EXEMPLARY			BELOW STANDARD
Connections to Standard and Learning Goals	④	③	②	①
Quality of Assessment Plan	④	③	②	①
Instructional Plan	④	③	②	①
Instructional Approaches	④	③	②	①
Curriculum Congruence	④	③	②	①
	TOTAL RATING			

PART C: EVALUATION SUMMARY

CURRICULUM MATERIAL	TOTAL RATING	MAJOR STRENGTHS	MAJOR WEAKNESSES	RECOMMENDATION

PART D: FINAL CURRICULUM EVALUATION TEAM RECOMMENDATIONS

SUMMARY REFLECTION

CHAPTER 6

1. After completing the seven steps in the Learning Path, describe your level of familiarity with the Central Ideas. Was your background adequate for completing the Curriculum Analysis?

2. How did your understanding of the principles of standards-based reform affect your ability to apply the Guide's procedure for analyzing and evaluating curriculum materials?

3. Describe any additional evaluative criteria that were added to the Curriculum Analysis Rubric? How did adding these descriptors affect the ratings that these materials received?

SUMMARY REFLECTION

CHAPTER 6

4. Did this standards-based procedure for completing a curriculum analysis and evaluation have merit as an initial screening process?

5. Elliott Eisner (1991) stated, "The ability to make fine-grained discriminations among complex and subtle qualities is an instance of what I have called connoisseurship" (p. 63). Describe the level of connoisseurship you reached in terms of your ability to evaluate standards-based instructional materials.

EXTENSION III: Curriculum Mapping

CHAPTER 7

Curriculum mapping is an increasingly popular approach for examining content area curriculum from a K-12 perspective. Typically, the principal goals of curriculum mapping are to ensure that the content being taught is appropriate for the grade level and to streamline instruction by filling gaps and avoiding unnecessary duplications. Because it has a K-12 orientation, curriculum mapping works best when conducted by district-wide teams of teachers. A school system that completes a curriculum mapping project can use these findings to adjust the level and frequency with which specific content is presented to students.

> *"There are too many examples of materials that include topics aligned with the content standards that do not address the fundamental understandings and abilities. Honorable mention is not enough to count as alignment of a curriculum with standards."*
>
> **Bybee, 1997**

When the curriculum for an instructional program has been modified according to a particular set of standards or curriculum framework, it is described as being aligned. In Extension III, users of the Guide utilize their knowledge of standards and learning goals to systematically apply a procedure for alignment.

EXTENSION III	LEARNING GOALS	ACTIVITIES
	Apply a systematic procedure for mapping a curriculum.	7.1: Preparing to Map a Curriculum
	Develop a K-12 curriculum framework that is aligned with the relevant standards.	7.2: Curriculum Mapping

SUMMARY REFLECTION	*3-2-1 Curriculum Mapping*

EXTENSION 7.1 Preparing to Map a Curriculum

Successfully mapping a curriculum requires the same level of background preparation as is needed for effectively analyzing and evaluating curriculum materials. In Chapter 6, we described a Learning Path through the Guide that would suitably prepare educators to systematically examine curriculum materials. Here we outline a five-step Learning Path that we recommend before a team begins an in-depth process of mapping a K-12 curriculum. However, if you have completed the Guide in its entirety, then a cursory review of the sections referenced in this Learning Path should provide sufficient advance preparation.

> *"The purpose of the (curriculum) task force is to review and monitor school curricula and ensure that every child in every school has the opportunity to meet the district's standards...schools have the flexibility to decide how to meet the standards, but they do not have the flexibility to decide whether to meet the standards."*
>
> **National Research Council, 1999**

K-W-L CURRICULUM MAPPING

1. Describe what you ***already know*** about Curriculum Mapping.

2. Make a list of things that you ***wonder*** about Curriculum Mapping.

3. After you complete Extension III, prepare a list of things you ***learned*** about Curriculum Mapping.

EXTENSION 7.1 Discovery Guide: Preparing for the Curriculum Mapping Learning Path

DIRECTIONS

The Curriculum Mapping Focusing Questions provide connections between the individual steps in the Learning Path and the templates that you will use in the process described in Extension 7.2. For each step, you should prepare a summary statement that states the Central Idea of the activity.

> *"Standards provide the structure from which a deep and rich local curriculum can be built."*
>
> **Carr & Harris, 2001**

STEP ONE

PREFACE TO THE GUIDE	Central Idea of Preface to the Guide:
Curriculum Mapping Focusing Question from Preface to the Guide: How are the four major premises of the Guide related to a comprehensive curriculum mapping initiative?	

STEP TWO

EXPLORATION 2.3	Getting to Know the National Standards Resources	Central Idea of Exploration 2.3:
Curriculum Mapping Focusing Question for Exploration 2.3: What information do the national documents provide about standards and learning goals that you can use during your curriculum mapping initiative?		

EXTENSION 7.1 Discovery Guide: Preparing for the Curriculum Mapping Learning Path

STEP THREE

EXPLANATION 3.1 — Content Clarification of a National Standard

Curriculum Mapping Focusing Question for Explanation 3.1:

How can a Content Clarification determine if the intent and key ideas behind a standard or learning goal are being thoroughly addressed in a curriculum?

Central Idea of Explanation 3.1:

STEP FOUR

EXPLANATION 3.2 — Creating a Content Crosswalk

Curriculum Mapping Focusing Question for Explanation 3.2:

Why is it important to check for consistency between the grade level placement of your state or local standards and the recommendations found in the national standards?

Central Idea of Explanation 3.2:

STEP FIVE

EXPLANATION 4.1 — Congruence With Standards

Curriculum Mapping Focusing Question for Explanation 4.1:

How does an understanding of congruence among standards, assessment, and instruction assist in making informed decisions about the placement of curriculum?

Central Idea of Explanation 4.1:

EXTENSION 7.2 Curriculum Mapping

> *"A key aspect of task analysis is the idea of aligning goals for learning what is taught, how it is taught, and how it is assessed."*
>
> **Bransford, Brown & Cocking, 1999**

School organizations typically create district-wide teams of teachers and administrators and use one of two major mapping approaches for analyzing their curriculum. Both procedures gather data about the existing curriculum and use these findings to make curricular adjustments. The two methods differ principally in what a district chooses to use as the target of its alignment. This could be a local curriculum framework or a set of accepted standards.

Some mapping teams begin by producing a curriculum framework that specifies a sequence for when certain teacher-selected topics or concepts should be addressed. These internal decisions about content placement are the ultimate basis for determining alignment. The next phase is to complete an audit of points across the K-12 spectrum where the desired topics are currently introduced. Comparisons are drawn between the recommendations contained in the desired curriculum framework and the actual grade levels where these topics are being taught. To achieve full alignment with this approach, topics are repositioned by grade level and gaps in the existing curriculum are filled.

The second type of mapping strategy begins with an established set of standards and learning goals for students as the starting point for your analysis. Standards provide the criteria against which an existing curriculum is compared. As the term *standard* implies, this approach to curriculum alignment emphasizes what students are expected to know or be able to do and by what grade levels (benchmarks) these learning goals should be met.

The Guide endorses a standards-based mapping approach because it reduces the likelihood that teachers will make curricular decisions based solely on tradition, omit important concepts, or act independently of the guiding principles found in the national standards. The procedure outlined in the Guide is called a match/gap analysis. It is similar to what has been used in the Council of Chief State School Officers (Tibbals, 2000) benchmarking process to analyze state standards documents. As the expression implies, *matches* occur when the standards agree with what and when something is currently being taught. *Gaps* are mismatches between what the standards recommend and what is actually occurring. Alignment is accomplished by filling gaps and moving curricula to more appropriate grade levels.

EXTENSION 7.2 Curriculum Mapping

Curriculum mapping projects achieve their greatest district-wide impact when completed by collaborative teams of K-12 teachers. For this reason, Extension III is written as a group activity. For curriculum mapping to be successful, it must be thoughtful and deliberate. Be sure to allocate sufficient time and resources. Mapping is a process that requires your team to (a) select a basis for determining alignment (we recommend using standards!), (b) collect information, and (c) use data as the basis for making decisions that are in the best interests of students.

DIRECTIONS

PART A: PREPARE A MAPPING TEMPLATE

Before beginning, you should create a mapping template for your relevant standards and benchmark them for all grade levels and courses. The State of Tennessee model (Figure 7.1 on page 124) is an example of a template that has been successfully used for a curriculum mapping project. Only a small portion of the full document is shown. You will notice that this mapping procedure focuses on learning goals for students as the basis for making decisions about alignment. As you study this Curriculum Mapping template, answer the following questions:

- Where are the standards and learning goals located on the template?
- Why is the Curriculum Mapping template organized by grade range/band?
- Where are the individual grade levels within a particular grade range listed?
- What is the purpose of the scoring variations?
- What data will eventually be placed in the individual boxes?
- How would you expand this partial template to make it usable for a K-12 mapping project?
- What modifications will be necessary to make the template more appropriate for working with your set of standards?

PART B: ORGANIZE THE MAPPING TEAM

Gathering preliminary data about the existing curriculum is best accomplished within smaller groups organized according to the grade clusters used in your standards document. It is advisable to create K-3, 4-5, 6-8, and 9-12 teams. These grade cluster teams can meet and have cross-cluster conversations as well. Ultimately, all of the findings will be compiled and examined from a K-12 perspective.

EXTENSION 7.2 Curriculum Mapping

PART C: STUDY THE SCORING VARIATIONS

Read the explanations for each variation. Discuss these carefully in your group before beginning. It is important that all members of the mapping team are in general agreement over the meaning and distinctions among the four variation levels.

VARIATION 4	*Curriculum thoroughly addresses the learning goal. Student performance is assessed.*	For you to assign this score to an individual grade level or course means that several lessons or activities specifically target this learning goal. Student performance is fully assessed for evidence that the learning goal has been met.
VARIATION 3	*Curriculum thoroughly addresses the learning goal. Student performance is not assessed.*	For you to assign this score to an individual grade level or course means that several lessons or activities specifically target this learning goal. However, student performance is not assessed for evidence that this learning goal has been met.
VARIATION 2	*Curriculum touches upon the learning goal. Student performance is not assessed.*	For you to assign this score to an individual grade level or course means that one or more lessons or activities touch upon the topic. However, student experiences neither specifically target nor assess this learning goal.
VARIATION 1	*Curriculum does not address the learning goal.*	For you to assign this score to an individual grade level or course means that topics or activities related to this learning goal are not introduced.

PART D: REVIEW YOUR CURRENT CURRICULUM ORGANIZATION

Remember, you must analyze your curriculum as it is presently being implemented to determine *if and how* each learning goal is targeted and/or assessed within your K-12 framework. A good procedure is to work from your mapping template. Review each learning goal carefully. If you are unsure as to the meaning or intent of this goal, consider completing a Content Clarification like you did in Chapter 3. Locate the grade level(s) at which this goal is actually addressed in your district.

Complete a careful analysis of each learning goal. In order to determine which scoring variation applies best, ask yourself the following questions:

EXTENSION 7.2 Curriculum Mapping

- Is it likely that students will achieve this learning goal through the curriculum and instruction that they are currently being provided? If the answer is yes, then the learning goal is thoroughly addressed.
- Do the lessons and activities directly target this learning goal, or are students merely being exposed to the topic suggested by the learning goal? If the answer is yes, then the learning goal is thoroughly addressed. If not, then the curriculum merely touches upon the learning goal. Reviewing the Content Clarification procedure (Explanation 3.1) may be necessary to understand the distinction between the topic connected to a standard and the deeper intent of a standard.
- Are the students' performance levels relative to this learning goal being assessed, and does this assessment provide valid evidence for the quality of my students' performance? If the answer is yes, then the learning goal is thoroughly assessed. You may want to revisit Chapter 4: Explanation II for a review of assessment principles.
- Finally, assign an appropriate score for each learning goal.

PART E: GATHER DATA

For learning goals that receive scores at *Variation levels 4 or 3*, you should provide specific information from your study by completing Figures 7.2-7.4: Curriculum Match Analysis (pages 125-127). For learning goals that receive scores at *Variation levels 2 or 1*, you should provide specific information from your study by completing Figures 7.5-7.7: Curriculum Gap/Placement Analysis (pages 128-130).

You may discover instances when the standard is thoroughly addressed but at an inappropriate grade level or addressed at more than one grade level. Special tables are provided for these cases.

PART F: SUMMARIZE FINDINGS

After you complete the curriculum mapping procedure, you should carefully analyze Figures 7.2-7.7 for all of the learning goals. Summarize your major alignment findings and prepare an action plan for achieving full K-12 alignment with the standards.

EXTENSION 7.2 Curriculum Mapping

PART G: DEVELOP AN ACTION PLAN

The mapping process answers the question: What are we currently doing? The answer to "What should we do next?" will determine how effectively you achieve K-12 alignment of your local curriculum. Unless and until the process of mapping a curriculum produces tangible outcomes, then the procedure is little more than an interesting professional development exercise.

Generating an Action Plan for achieving full alignment of your curriculum with the set of standards used for the Curriculum Mapping procedure is strongly recommended. The following action framework presented in Figure 7.8 on page 131 was adapted from the National Staff Development Council (2001) and is one that you may find useful. Carr and Harris (2001) offer other detailed strategies for developing a comprehensive action plan aimed at systemwide implementation of standards.

Congratulations! You have just taken an important first step toward implementing a standards-based curriculum. Please think about your experiences as you mapped your curriculum. Then, take a few moments to complete the summary reflection.

FIGURE 7.1 TENNESSEE'S CURRICULUM MAPPING TEMPLATE

Scoring Variations:

Four: Curriculum thoroughly addresses the LG. Student performance is assessed.
Three: Curriculum thoroughly addresses the LG. Student performance is not assessed.
Two: Curriculum touches upon the LG. Student performance is not assessed.
One: Curriculum does not address the LG.

	Life Science Standards and Learning Goals—By the end of Grade 3, students...	K	1st	2nd	3rd
1.0	Cell Structure and Function				
1.1	Recognize that living things are made up of smaller parts				
1.2	Recognize that smaller parts of living things contribute to the operation and well-being of entire organisms				
2.0	Interactions Between Living Things and Their Environment				
2.1	Recognize the distinction between living and non-living things				
2.2	Realize that organisms use their senses to interact with their environment				
2.3	Examine interrelationships among plants, animals, and their environment				
2.4	Recognize that the environment and the organisms that live in it can be affected by pollution				
3.0	Food Production and Energy for Life				
3.1	Recognize the basic requirements of all living things				
3.2	Recognize the basic parts of plants				
4.0	Heredity and Reproduction				
4.1	Recognize that living things reproduce				
4.2	Recognize that offspring tend to resemble their parents				
4.3	Recognize that the appearance of plants and animals change as they mature				
5.0	Diversity and Adaptation Among Living Things				
5.1	Recognize the differences among plants and animals of the same kind				
5.2	Recognize that living things have features that help them to survive in different environments				
6.0	Biological Change				
6.1	Recognize that some animals that once lived are no longer found on earth				

FIGURE 7.2 CURRICULUM MATCH ANALYSIS (SCORE OF 4)

MATCH: Curriculum thoroughly addresses this Learning Goal.
Student performance is assessed.

Grade Level/Course	Content Area	LG Number

Learning Goal	Variation Score
	4

Learning opportunities for students (lessons and activities) that target this Learning Goal	How student performance is assessed

FIGURE 7.3 CURRICULUM DUPLICATION ANALYSIS (SCORE OF 4)

DUPLICATION: Curriculum thoroughly addresses and assesses this Learning Goal at multiple grade levels

Grade Levels/Courses	Content Area	LG Number

Learning Goal	Variation Score
	4

Learning opportunities for students (lessons and activities) that target this Learning Goal	How student performance is assessed

Proposed Actions
TASKS:
IMPLEMENTATION TIME LINE:
NECESSARY RESOURCES:

FIGURE 7.4 CURRICULUM MATCH ANALYSIS (SCORE OF 3)

MATCH: Curriculum thoroughly addresses this Learning Goal.
Student performance is ***not*** assessed.

Grade Level/Course	Content Area	LG Number
Learning Goal		Variation Score
		3
Learning opportunities for students (lessons and activities) that target this Learning Goal		
Proposed Actions		
TASKS:		
IMPLEMENTATION TIME LINE:		
NECESSARY RESOURCES:		

FIGURE 7.5 CURRICULUM PARTIAL MATCH ANALYSIS (SCORE OF 2)

PARTIAL MATCH: Curriculum touches upon this Learning Goal.
Student performance is ***not*** assessed.

Grade Level/Course	Content Area	LG Number
Learning Goal		Variation Score
		2
Describe contact with the Learning Goal		
Proposed Actions		
TASKS:		
IMPLEMENTATION TIME LINE:		
NECESSARY RESOURCES:		

FIGURE 7.6 CURRICULUM GRADE PLACEMENT ANALYSIS (SCORE OF 2)

PARTIAL MATCH: Curriculum addresses this Learning Goal, but at an inappropriate grade level.

Grade Level/Course	Content Area	LG Number
Learning Goal		Variation Score
		2
Suggested Grade Placement		
Proposed Actions		
TASKS:		
IMPLEMENTATION TIME LINE:		
NECESSARY RESOURCES:		

FIGURE 7.7 CURRICULUM GAP ANALYSIS (SCORE OF 1)

GAP: Curriculum does not address this Learning Goal.

LG Number	Learning Goal	Variation Score
		1

Most appropriate approach for targeting this Learning Goal

Proposed Actions
TASKS:
IMPLEMENTATION TIME LINE:
NECESSARY RESOURCES:

FIGURE 7.8 CURRICULUM MAPPING ACTION FRAMEWORK

Separate the principal findings according to gaps, duplications, and grade misplacements.

- **Prepare** a summary of major tasks
- **Generate** a list of action steps
- **Identify** the potential obstacles to full alignment
- **Develop** an implementation time line
- **Speculate** on the implications of achieving alignment
- **Identify** the necessary resources

SUMMARY REFLECTION

CHAPTER 7

3-2-1 CURRICULUM MAPPING

A. What ***three big ideas*** emerged about how curriculum mapping can help to align a K-12 curriculum with a set of learning goals for students?

> *"Ultimately, curriculum mapping should produce a seamless curriculum for a K-12 district that primarily benefits students."*
>
> **DeClark, 2002**

3
2
1

B. What ***two immediate actions*** do you plan to take as a result of having mapped your curriculum?

2
1

C. What ***one major insight*** did you gain from mapping your curriculum?

1

EVALUATION: Discoveries About Standards-Based Teaching and Learning

CHAPTER 8

> *"Ideas acquired with ease are discarded with ease."*
>
> **Pascale, 1990**

> *"Reflection gets to the heart of the matter, the truth of things. After appropriate reflection, the meaning of the past is known, and the resolution of the experience—the course of action you must take as a result—becomes clear."*
>
> **Bennis, 1994**

The purpose of the evaluation stage of a learning cycle is to examine changes in thinking that can be directly attributed to specific learning experiences. In an evaluation, learners reflect on an activity to uncover and clarify ideas that have emerged and concepts that have changed. Through this type of self-analysis, new information becomes assimilated into a person's pre-existing knowledge base.

Like our fictional teacher, Julian, introduced in Chapter 4, working through the Guide required you to expend significant amounts of time, energy, and careful thought. You completed a journey that included careful consideration of standards and reviews of the national documents that support the reform movement. You investigated how assessment could be used to drive the selection of learning experiences for students. You applied standards-based principles to develop and evaluate curriculum materials. Some may have used these same ideas to map your curriculum across grade levels. The Guide hopefully gave you the tools necessary for implementing the central features of standards-based curriculum and instruction in your classroom.

In Chapter 8, you have a final opportunity to reexamine experiences that have hopefully led to transformations in your understanding of concepts that lie at the core of the Guide.

EVALUATION	LEARNING GOALS	ACTIVITIES
	Analyze changes in beliefs and understanding	8.1: Understandings About Constructivism and Standards-Based Learning
	Assess what is discovered by completing each major section of the Guide	8.2: National Standards as Resources for Instruction
		8.3: Assessment That Guides Teaching and Learning
		8.4: Developing and Evaluating Standards-Based Materials
	Evaluate the overall impact of the Guide	8.5: Standards in the Classroom: An Implementation Guide

EVALUATION 8.1 Understandings About Constructivism and Standards-Based Learning

In this reflection, you revisit Engagement 1.2 and make a pre/post comparison of your initial and final thoughts. The differences between the two sets of results should indicate changes that are connected with your experiences using the Guide.

> *"Teachers get information thrown at them and are expected to just work through it."*
>
> **Past user of the Guide**

DIRECTIONS

The first time that you completed this survey, you examined your initial attitudes and beliefs about standards and constructivism. Place an X where you now fall on the scale that contrasts the characteristics of constructivist, standards-based learning environments with more traditional classrooms. Use an O to mark the spot where you classified yourself as you began using the Guide.

EXAMPLE:

Emphasis on manipulative curriculum materials	............X...............................O..	Emphasis on text materials

1	Varied instructional approaches	....	Lecture as principal method of instruction
2	Teacher as mediator of learning environment	....	Teacher as disseminator of information
3	Teacher working with groups of students	....	Teacher generally at the front of the class
4	Collaboration among teachers	....	Teachers working in isolation
5	Support through ongoing professional development	....	Professional development a personal initiative
6	High standards for all students	....	High standards for the best and brightest
7	High expectations for all students	....	Different expectations for different students
8	Not all standards reached by all students at same rate	....	Expected progress identical for all students
9	Prior knowledge of student critical to learning	....	Students' minds are blank slates
10	Students as active learners	....	Students as recipients of information

EVALUATION 8.1 Understandings About Constructivism and Standards-Based Learning

11	Collaboration among students	....	Students working alone
12	Students assume greater responsibility of work	....	Teacher is the worker
13	Student questions highly valued	....	Questions provided by teacher
14	Curriculum decisions based on standards	....	Curriculum driven by coverage of text material
15	Emphasis on student learning	....	Emphasis on teaching
16	Focus on understanding	....	Focus on acquiring factual knowledge
17	Clear expectations for students	....	Expectations for students not openly stated
18	In-depth study of fewer topics	....	Coverage is broad and lacks depth
19	Integration of content areas	....	Content areas separate
20	Experimental, inquiry-based learning	....	Teaching driven by texts and workbooks
21	Curriculum adapted for individual students	....	One curriculum for all
22	Multiple sources of evidence for student learning	....	Few sources of evidence for student learning
23	Continuous monitoring of student progress	....	End-of-unit tests
24	Numerous opportunities for self-assessment	....	Little self-assessment
25	Student progress reported in terms of standards	....	Traditional grade reporting
26	Heterogeneously grouped classes	....	Ability tracking of students
27	Student-centered instruction	....	Teacher-centered instruction
28	Students guided toward inquiry	....	Lock-step approaches

REFLECTION 8.1 Understandings About Constructivism and Standards-Based Learning

Here are some final reflection questions about constructivism and standards for you to consider.

> *"I can now see the research that went into developing the national standards and recognize how they impacted the various state standards."*
>
> **Past user of the Guide**

1. What are the major changes between your initial and final responses?

2. What specific activities in the Guide do you associate with the areas of greatest change?

3. After completing the Guide, how would you describe your present state of readiness for moving toward a standards-based and constructivist style of teaching? What influence, if any, has the Guide had on your teaching?

REFLECTION 8.1 Understandings About Constructivism and Standards-Based Learning

4. What do you now regard as the major challenges and obstacles to implementing standards-based instruction?

5. How would you describe your current level of confidence for overcoming these above listed obstacles?

EVALUATION 8.2 National Standards as Resources for Instruction

One of the original premises of the Guide was that learning how to use standards as the basis for curriculum and instruction is challenging. Therefore, teachers must be provided with adequate resources before they can successfully adopt the principles of standards-based reform. We believe that the national standards documents have been an underutilized resource in this regard.

> *"The best audience for the Guide consists of teachers who want their children to succeed and meet all of the benchmarks for that grade level."*
>
> **Past user of the Guide**

The Content Crosswalk and Content Clarification provide the keystones for the Guide. Together, they illustrate how the national standards documents contribute toward helping teachers gain a deeper understanding about standards and their connection to ideas concerned with content, learning, and teaching. These sentence stems are designed to assist you in reflecting upon how the national standards documents have influenced your thinking about professional practice. This evaluation references Chapters 2 and 3 of the Guide.

The national standards documents are …	
Before using the Guide, my familiarity with the national standards documents was …	
After using the Guide, my familiarity with the national standards documents is …	
For me, the principal value of the national standards documents is …	
Completing a Content Crosswalk illustrated that …	
Completing a Content Clarification enabled me to …	
The major difference between my initial and present feelings about national standards is…	

EVALUATION 8.3 Assessment That Guides Teaching and Learning

According to the North Central Regional Educational Laboratory (NCREL), its Nine Assessment Principles listed below provide a foundation for effectively making judgments about the quality of students' performance. They are provided as a reference. Completing the Assessment Principles Survey (Figure 8.1) should help you to characterize your current understanding of student assessment as you finalize your work with the Guide. This evaluation extends the ideas introduced in Chapter 4: Explanation I.

> *"I feel more confident about organizing teaching experiences with the standard and performance assessment up front."*
>
> **Past user of the Guide**

NCREL Nine Assessment Principles

1. Regular assessment of student progress and achievement is part of good teaching.
2. Assessment tasks must be presented in a way so that the student is perfectly clear about what is expected, and grades or marks awarded.
3. Assessment tasks should be designed so that most children in a group do well on most tasks.
4. Children should only be assessed on knowledge, skills, and attitudes their teacher has given them opportunities to develop.
5. A combination of different methods is vital if we are to get a balanced picture of student development.
6. The main purpose of assessment is to help students learn.
7. Children must be fluent in the language in which they are assessed, and the level of the language used must match their stage of development.
8. Students themselves can often be asked to assess their own level of achievement and the achievements of their classmates.
9. Assessment should focus on each student's achievements, independently of how other students are doing.

FIGURE 8.1 NCREL ASSESSMENT PRINCIPLES SURVEY

Consider these statements below and rate your "agreement" with each. Choose "Agree Completely" if the principle expressed affirms your own beliefs. If any principles challenge your own beliefs or experiences, choose "Disagree Completely." Individually record your ratings on this survey form.	AGREE CCOMPLETELY	AGREE MORE THAN DISAGREE	DISAGREE MORE THAN AGREE	DISAGREE COMPLETELY	HAVE NO OPINION
1. Learners need to find out how well they are doing and teachers need to find out how successfully they are teaching. Therefore, regular assessment of student progress and achievement is part of good teaching.	O	O	O	O	O
2. The main purpose of assessment is to help students learn. When students are assessed well and given feedback about their performance, they find out what they have learned successfully and what they have not. Any weaknesses can then be reduced.	O	O	O	O	O
3. Assessment tasks should be designed so that most children in a group do well on most tasks. This takes the threat out of being assessed and allows children to be motivated to learn by the regular experience of success and praise.	O	O	O	O	O
4. Design/selection of assessment tasks requires a clear idea of the curriculum objectives. Children should only be assessed on knowledge, skills, and attitudes their teacher has given them opportunities to develop, and each task should be well within the capabilities of most students.	O	O	O	O	O
5. No single method of assessment can give information about achievement of the full range of learning objectives. Therefore, a combination of different methods is vital if we are to get a balanced picture of student development.	O	O	O	O	O
6. Assessment tasks must be presented in a way so that the student is perfectly clear about what is expected, and grades or marks must be awarded so that the student feels s/he has been fairly treated.	O	O	O	O	O
7. The language of assessment must match the language of instruction. If not; then assessment produces unfair and invalid results. Children must be fluent in the language in which they are assessed, and the level of language used must match their stage of development.	O	O	O	O	O
8. The teacher's unbiased judgments are important in assessment, but students themselves can often be asked to assess their own level of achievement and the achievement of their classmates. They can be (surprisingly) accurate and honest.	O	O	O	O	O
9. Assessment should focus on each student's achievements, independently of how other students are doing. Constant comparison/competition with classmates can damage the self-esteem and self-confidence of many students.	O	O	O	O	O
10. Alternative assessment is not necessarily better than "traditional" assessment.	O	O	O	O	O
11. Assessment done "in the service of instruction"—not merely as a tool for monitoring progress or evaluating achievement—is an important resource for learning.	O	O	O	O	O
12. There must be no secrets in assessment: Students need to know and internalize the expectations and criteria for good performance.	O	O	O	O	O
13. Students should be involved in setting goals and establishing the criteria.	O	O	O	O	O
14. Student responses should be scored according to specified criteria, known in advance, which define standards of good performance.	O	O	O	O	O
15. The assessment task itself must measure meaningful instructional activity, with most of the class time devoted to it.	O	O	O	O	O
16. The knowledge, skills, and habits of mind assessed should match the teacher's educational objectives and instructional emphases.	O	O	O	O	O
17. Students should be performing, creating, producing, or doing some thinking about just what it is they're performing, creating, producing, and doing.	O	O	O	O	O
18. Tasks should require students to use higher-order thinking or problem-solving skills.	O	O	O	O	O
19. It is as important to assess a student's opportunity to learn as it is to assess a student's learning.	O	O	O	O	O

REFLECTION 8.3 Assessment That Guides Teaching and Learning

3-2-1 ASSESSMENT

> "There is so much that needs to be linked. People need to take standards one step at a time."
>
> **Past user of the Guide**

A. After completing the Guide, what are the ***three major changes*** in your views and understandings about assessment?

3
2
1

B. What are the ***two major challenges*** that you are likely to encounter in implementing your new knowledge of assessment?

2
1

C. What is the ***one key insight*** that you gained about student assessment?

1

EVALUATION 8.4 Developing and Evaluating Curriculum Materials

> *"Beginning with a standard...what a concept! Again so simple! I've always plugged them in at the end ...definitely not standards-based teaching and learning!"*
>
> **Past user of the Guide**

Curriculum development and effective teaching rely upon a different subset of skills. Most skillful teachers lack the time, energy, or opportunity to make serious attempts at developing original curriculum materials. Yet these same teachers can quickly spot high-quality materials and find clever ways to weave them into the fabric of their courses. Huberman (1995) refers to this constant tinkering with curriculum as *bricolage*. Teachers "pick up a new technique here, a new activity there, and materials that seem to fit their own styles, settings and students, then adjust them on the basis of their goals and experience" (Thompson & Zeuli, 1997, p. 14). It is in this way that teachers generally "develop" curriculum.

Teachers often require assistance to complete formal analyses and evaluations of existing curricula based on contemporary ideas about standards, alignment, and assessment. Chapter 6 introduced a systematic process for selecting curriculum materials based on principles embedded in the Guide. The following PMI Summary Reflections (de Bono, 1982) reference Chapters 5 and 6 of the Guide.

Pluses of developing your own curriculum materials +	
Minuses of developing your own curriculum materials −	
Interesting parts of developing your own curriculum materials !	

Pluses of evaluating existing curriculum materials +	
Minuses of evaluating existing curriculum materials −	
Interesting parts of evaluating existing curriculum materials !	

EVALUATION 8.5 Standards in the Classroom: An Implementation Guide

Our major purpose in developing the Guide was to help teachers become better informed practitioners of standards-based principles in their classrooms. Here is an opportunity to capture some final thoughts before beginning to apply your new understandings and beliefs about standards as the basis for teaching and learning.

> *"You'll value what students learn if it ties into the standards."*
>
> **Past user of the Guide**

The reason I initially decided to use the Guide was because …	
When I began using the Guide, my first impressions were …	
As a result of using the Guide, my beliefs about standards …	
As a result of using the Guide, my understanding of standards …	
After completing the Guide, I now feel more confident about …	
After completing the Guide, I am still a bit uneasy about …	
For me, the major drawback to using the Guide was …	
The major benefit that I gained from using the Guide is …	
If a colleague asked me about the Guide, I would say …	

The authors would appreciate receiving your feedback and suggestions for improvement for inclusion in future editions of the Guide. Please e-mail your final reflection.

Richard Audet, Ed.D.
Roger Williams University
One Old Ferry Road
Bristol, RI 02809
raudet@rwu.edu

Linda K. Jordan, Ed.S.
Tennessee Department of Education
710 James Robertson Parkway
Nashville, TN 37243-8536
linda.k.jordan@state.tn.us

ENDURANCE: Sustaining Change

CHAPTER 9

After most of the barriers to change have been overcome and professional development has been provided through the Guide, what additional measures should be taken to ensure that change is implemented and sustained?

> *"The more things change, the more they remain the same."*
>
> **Janvier, 1849**

Revisiting two of the Guide's original premises makes the case for supporting changes that occurred as a result of your experiences. One idea was that "standards have ramifications that permeate the entire educational system." Thus, everyone who is engaged in helping people to learn has both personal and broadly defined professional stakes in reform. The second principle was that "learning about standards is neither simple nor direct." Translating standards into practice requires a major transformation in the way that educators view teaching and learning and thus a concomitant need for ongoing professional development.

Much of the original research about individual behavioral change comes from studies of smokers. Researchers found a consistent pathway of stages ranging from initial contemplation to eventual termination. These patterns have been generalized across other areas of personal change, including teaching. A core finding of this work is captured by the maxim, "Change is a process, not an event."

Let's begin with a question. When talking about change in educational settings, is it best to conceptualize the process from an individual or a group perspective? Because the wave of educational reform is systemic in nature, the current focus in teaching and learning circles is on organizational change. However, a renowned thinker on the subject of change, Peter Senge (1990), notes that, "In human systems, people often have potential leverage that they do not exercise because they focus only on their own decisions and ignore how their decisions affect others" (p. 40). Senge's thoughts provide a framework for reconciling the apparent dichotomy between changes as they occur in individuals and institutional change, that is, piecemeal versus wholesale.

This Guide focuses on teacher change within a standards-based system. The answer to the perspective on change question asked earlier is, both. The best way for an individual to sustain change is as a member of a larger community that is committed to systemic reform. This chapter describes one approach for building upon the experiences gained from the Guide.

REFLECTION

CHAPTER 9

Most contemporary images of professional teaching careers attach major importance to sustained professional development. These continuous improvement models argue that change is best initiated and supported through collaborative inquiry within communities of professionals. Sustaining transformative change requires a long-term commitment from teachers and administrators.

> *"Teachers react against ideas and materials that are theoretically sound but do not function in the classroom. They seek proof that other professionals with whom they identify are making new methods work...the fact that others can do it gives them moral support and challenges them."*
>
> **Black & Atkin, 1996**

Before considering strategies for accomplishing the goal of sustaining change, complete the following discovery guide. Describe your reaction to each statement in terms of your current attitudes, beliefs, and understandings about implementing standards in the classroom.

1. In educational circles, trends come and go but things tend to remain as they were.	
2. An educational system will continue to move in a straight line unless something like a commitment to standards forces it to change direction.	
3. Applying a continuous force will result in continuous change.	
4. Small forces can cause great changes in the direction of a system over time.	
5. The only time that a changing system grinds to a halt is when something forces it to stop.	

Adapted from Carr, J. F. & Harris, D. E. (2001). *Succeeding with standards*. Alexandria, VA: Association for Supervision and Curriculum Development. p. 146.

STRATEGIC PLAN FOR SUSTAINING CHANGE

CHAPTER 9

Among the principles of effective professional development identified by Darling-Hammond and McLaughlin (1995) and Loucks-Horsley et al. (1998) are that it be directly connected with a teacher's work with students, involve problem solving and experimentation, and include collaboration among teachers. Based on these three characteristics of high-quality professional development, we offer a three-pronged approach for sustaining your individual efforts to implement standards in your classroom and as a member of a broader community of reform (Figure 9.1). Study groups offer the context for sustaining change; action research provides the process. Standards-based curriculum and instruction are the two objects of professional inquiry into practice.

FIGURE 9.1 VISUALIZATION FOR SUSTAINING CHANGE

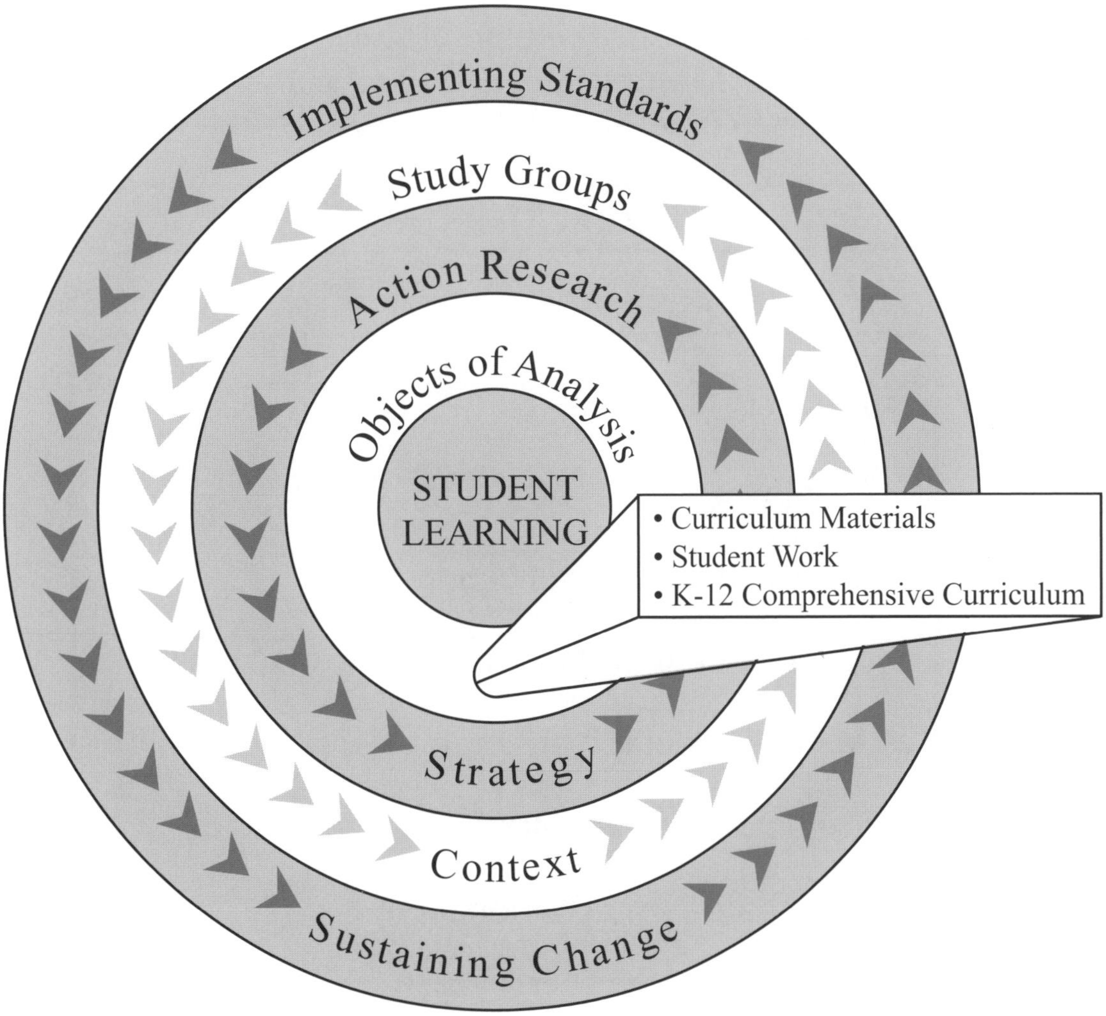

STRATEGIC PLAN FOR SUSTAINING CHANGE

PART A

STUDY GROUPS: A CONTEXT FOR SUSTAINING CHANGE

Earlier models of professional development begin with a faculty needs assessment followed by the identification of qualified consultants who can provide the desired services. Expertise generally resides outside of the system.

> *"Because the (whole-faculty study group) model evolves from how teachers actually work together in schools, it is fluid, flowing and readjusting."*
>
> **Murphy & Lick, 2001**

Prevailing ideas about effective professional development (National Staff Development Council, 2000a) hold that expertise can be developed and shared within a community of practice. An increasingly popular context for supporting this type of professional development is through the formation of study groups. Such vehicles for professional growth are embedded in the actual classroom practice of participating teachers and are centered on learners. Although change is generally the desired outcome of a study group, such structures are highly appropriate for supporting implementation of new ideas and sustaining ongoing changes in practice, such as a transition to standards-based instruction.

FIGURE 9.2 CHARACTERISTICS OF STUDY GROUPS

- Are based on the needs of students
- Employ a process of inquiry
- Inspire individual reflection
- Are anchored in group members' own practice
- Apply democratic principles
- Promote construction of new knowledge
- Have built-in motivational factors

STRATEGIC PLAN FOR SUSTAINING CHANGE

PART B

ACTION RESEARCH: A STRATEGY FOR SUSTAINING CHANGE

For study groups to progress and answer questions, they need tools for inquiry. One professional development approach that has blossomed during recent years is action research. The process begins with formulating questions about issues that concern teachers. Systematic, but informal investigations aimed at generating qualitative data follow. The information generated is generally aimed at effecting changes in teaching and learning.

> *"Undertaking research in ... classrooms is one way in which teachers can take increased responsibility for their actions and create a more energetic and dynamic environment in which teaching and learning can occur."*
>
> **Hopkins, 1992**

There is a large body of literature accessible via the Internet that can provide specific guidance for your group. In fact, investigating the topic of action research itself might be a nice project for launching the study group. As you explore the Internet, you can use the following Discovery Guide to organize your findings.

DIRECTIONS

Use this framework for organizing your information about action research.

Characteristics of Action Research	
Qualitative vs. Quantitative Research	
Role of Questions in Action Research	
Common Strategies for Conducting Action Research	
Sources of Data	
Approaches for Reporting Conclusions	
Moving From Research to Action	

STRATEGIC PLAN FOR SUSTAINING CHANGE

PART C

POTENTIAL TOPICS FOR INVESTIGATION

> *"Action research is a systematic process for investigating an authentic school or classroom situation in order to improve actions or instruction. It is preplanned, organized, and should be shared."*
>
> **Johnson, 2002**

In a learning cycle, extensions provide opportunities for learners to apply their new understandings in different situations. This approach is consistent with cognitive flexibility theory (Spiro, Coulson, Feltovich, & Anderson, 1992) which regards the ability to transfer knowledge beyond the original learning experience as the best indicator of effective learning. This theory is a constructivist belief rooted in the idea that individuals develop their own representations of knowledge that they can apply adaptively to the particular demands of new situations.

From this perspective, the three extensions in the Guide provide an ideal context for action research projects. Here are some questions that may provide a focus for investigations within your study group.

- What can looking at assessments and related student work tell us about standards-based performances? (Chapter 4)
- How can existing curriculum materials be modified to incorporate the principles of standards-based design? (Chapter 5)
- How can the findings from examining student work be used to develop more effective standards-based learning experiences for students? (Chapter 5)
- How does the implementation of a standards-based curriculum affect student performance? (Chapter 6)
- What is the impact of implementing a strategic action plan that aligns a comprehensive K-12 curriculum with standards? (Chapter 7)

These questions are simply food for thought. As your study group matures, many ideas are sure to emerge that you will want to investigate. Good luck!

RESOURCES AND FURTHER READINGS

CHAPTER 10

"In the new view of professional development, teachers are engaged in professional learning every day, all day long."

National Staff Development Council, 2000a

LEARNING CYCLE

Biological Sciences Curriculum Study. (1993). *Developing biological literacy*. Dubuque, IA: Kendall Hunt.

Lawson, A. E., Abraham, M. R., & Renner, J. W. (1989). *A theory of instruction: Using the learning cycle in teaching science concepts and thinking skills*. Monograph of the National Association for Research in Science Teaching, 1. Cincinnati: NARST.

Llewellyn, D. (2002). *Inquire within: Implementing inquiry-based science standards*. Thousand Oaks, CA: Corwin.

Marek, E. A. & Cavallo, A. M. (1997). *The learning cycle*. Portsmouth, NH: Heinemann.

Spiro, R. J., Coulson, R. L., Feltovich, P. J. & Anderson, D. (1992). Cognitive flexibility, constructivism and hypertext: Random access instruction for advance knowledge acquisition in ill-structured domains. In T. Duffy & D. Jonassen (Eds.), *Constructivism and the technology of instruction*. Hillsdale, NJ: Erlbaum.

TEACHER CHANGE

Bellanca, J. (1998). Teaching for intelligence. *Phi Delta Kappan*, May, 658-660.

Darling-Hammond, L. & McLaughlin, M. (1995). Policies that support professional development in an era of reform. *Phi Delta Kappan, 76,* 597-604.

Fullan, M. G. (1993). *Change forces*. New York, NY: Falmer.

Fullan, M. G. & Hargreaves, A. (2000). There can be no improvement without the teacher. *encFocus, 7,* 19.

Guskey, T. R. (1986). Staff development and the process of teacher change. *Educational Researcher, 15,* 5, 5-12.

Hord, S. M., Rutherford, W. L., Huling-Austin, L., & Hall, G. G. (1987). *Taking charge of change*. Austin, TX: Southeast Educational Development Laboratory.

Pajeres, M. F. (1992). Teachers' beliefs and educational research: Cleaning up a messy construct. *Review of Educational Research, 62*, 307-332.

Richardson, V. (1990). Significant and worthwhile change in teaching practice. *Educational Researcher, 19*, 1, 10-18.

Senge, P. (1990). *The fifth discipline*. New York: Doubleday.

Sparks, D. (2001). Change: It's a matter of life or slow death. *Journal of Staff Development, Fall*, 49 - 53.

STANDARDS

American Association for the Advancement of Science. (1989). *Science for all Americans*. New York: Oxford University Press.

American Association for the Advancement of Science. (1993). *Benchmarks for science literacy*. New York: Oxford University Press.

Anderson, R. D. & Helms, J. V. (2001). The ideal of standards and the reality of schools: Needed research. *Journal of Research in Science Teaching, 38*, 1, 3-16.

Biological Science Curriculum Study. (1993). *Developing biological literacy*. Dubuque, IA: Kendall/Hunt Publishing.

Bybee, R. W. (1997). *Achieving scientific literacy*. Portsmouth, NH: Heinemann.

Driver, R., Squires, A., Rushworth, P. & Wood, V. (1994). *Making sense of secondary science.* London: Routledge.

Foriska, T. J. (1998). *Restructuring around standards*. Thousand Oaks, CA: Corwin.

Keeley, P. (2001). *A leader's guide to bridging Maine's learning results with national science standards*. Augusta, ME: Maine Mathematics and Science Alliance.

Kendall, J. S. & Marzano, R. J. (1997). *Content knowledge: A compendium of standards and benchmarks for K-12 education*. Aurora, CO: Mid-Continent Educational Laboratory.

Kennedy, M. M. (1998). Education reform and subject matter knowledge. *Journal of Research in Science Teaching, 35*, 3, 249-263.

Kohn, A. (2000). *The case against standardized testing*. Portsmouth, NH: Heinemann.

Leonard, W., Penick, J., & Douglas, R. (2002). What does it mean to be standards-based? *The Science Teacher*, *April,* 36-39.

Lynch, S. (1997). Novice teachers' encounter with the national science education reform: Entanglements or intelligent interconnections? *Journal of Research in Science Teaching, 34,* 1, 3-17.

National Commission on Teaching and America's Future. (1996). *What matters most: Teaching for America's future*. New York: Author.

National Council of Teachers of Mathematics. (2000). *Principles and standards for school mathematics*. Reston, VA: Author.

National Research Council. (1996). *National science education standards.* Washington, DC: National Academy Press.

National Research Council. (2002). *Investigating the influence of standards.* Washington, DC: National Academy Press.

Reeves, D. B. (1998). *Making standards work.* Denver, CO: Center for Performance Assessment.

Schmoker, S. & Marzano, R. J. (1999). Realizing the promise of standards-based education. *Association for Supervision and Curriculum Development, 56*, 6, 17-22.

Solomon, P. G. (1998). *The curriculum bridge: From standards to actual classroom practice*. Thousand Oaks, CA: Corwin.

Tucker, M. S. & Codding, J. B. (1998). *Standards for our schools.* San Francisco, CA: Jossey-Bass.

U.S. Department of Education. (1991). *America 2000: An educational strategy.* Washington, DC: Author.

ASSESSMENT

Arter, J. & McTighe, J. (2000). *Scoring rubrics in the classroom.* Thousand Oaks, CA: Corwin.

Black, P. & Atkin, M. J. (1996). *Charging the subject: Innovations in science, mathematics, and technology education.* London: Routledge.

Bransford, J. D., Brown, A. L., & Cocking, R. R. (1999). *How people learn.* Washington, DC: National Research Council.

Brown, A. L. (1994). The advancement of learning. *Educational Researcher*, *23*, 4-12.

Brown, J. H. & Shavelson, R. J. (1996). *Assessing hands-on science.* Thousand Oaks, CA: Corwin.

Damian, C. (2000). Facing and embracing the assessment challenge. *encFocus, 7,* 2, 16-17.

DO-IT. (2000). *Accommodating students with disabilities in math and science* [Videotape]. (Available from the University of Washington, Seattle)

Doran, R., Chan, F., & Tamir, P. (1998). Science educator's guide to assessment. Arlington, VA: NSTA Press.

Hammerman, E. & Musial, D. (1995). *Classroom 2061: Activity-based assessment in science.* Arlington Heights, IL: Skylight.

Hart, D. (1994). *Authentic assessment.* Reading, MA: Addison Wesley.

Hein, G. E. & Price, S. (1994). *Active assessment for active science.* Portsmouth, NH: Heinemann.

Hibbard, K. M. (2000). *Performance-based learning and assessment in middle school science.* Poughkeepsie, NY: Eye on Education.

Madfes, T. J. & Muench, A. (1999). *Learning from assessment.* San Francisco, CA: WestEd.

National Council of Teachers of Mathematics. (1995). *Assessment standards for school mathematics.* Reston, VA: Author.

National Research Council. (2001). *Classroom assessment and the National Science Education Standards*. Washington, DC: National Academy Press.

National Research Council. (2001). *Knowing what students know: The science and design of educational assessment.* Washington, DC: National Academy Press.

Reeves, D. B. (2001). *Performance assessment series: Elementary school edition*. Denver, CO: Advance Learning Press.

Reeves, D. B. (2001). *Performance assessment series: Middle school edition.* Denver, CO: Advance Learning Press.

Reynolds, D., Doran, R., Allers, R. & Agruso, S. (1996). *Alternative assessments in science: A teacher's guide*. Buffalo, NY: University of Buffalo.

Stigler, J. W. & Hiebert, J. (1999). *The teaching gap.* New York: Free Press.

CURRICULUM AND INSTRUCTION

American Association for the Advancement of Science. (2000). *Designs for science literacy.* New York: Oxford University Press.

American Association for the Advancement of Science. (2001). *Atlas of science literacy.* Washington, DC: American Association for the Advancement of Science and National Science Teachers Association.

Ausubel, D. (1968). *Educational psychology: A cognitive view.* New York: Holt, Rinehart, and Winston.

Covey, S. (1989). *The seven habits of highly effective people.* New York: Fireside.

Glatthorn, A. C. (1999). *Performance standards and authentic learning.* Larchmont, NY: Eye on Education.

Harris, D. E. & Carr, J. F. (1996). *How to use standards in the classroom.* Alexandria, VA: Association for Supervision and Curriculum Development.

Holloway, J. H. (2001). The use and misuse of standardized tests. *Educational Researcher, 20,* 4, 77-78.

Huberman, M. (1995). Networks that alter teaching: Conceptualizations, exchanges, and experiments. *Teachers and Teaching: Theory and Practice, 1,* 193-211.

Hyerle, D. (1996). *Visual tools for constructing knowledge.* Alexandria, VA: Association for Supervision and Curriculum Development.

Renaissance Group. (1999). *Inclusion: Children who learn together, live together.* Retrieved from http://www.uni.edu/coe/inclusion.

Wiggins, G. & McTighe, J. (1998). *Understanding by design.* Alexandria, VA: Association for Supervision and Curriculum Development.

CURRICULUM ANALYSIS

Biological Sciences Curriculum Study. (2002). *Analyzing instructional materials (AIM) process.* Colorado Springs, CO.

Bush, W. S., Kulm, G. & Surati, D. (2000). Getting together over a good book. *Journal of Staff Development*, Spring, 35- 38.

Bybee, R. W. (Ed.). (1997). *Learning science and the science of learning: Science educator's essay collection.* (pp. 121-136). Alexandria, VA: National Science Teachers Association Press.

Eisner, E. (1991). *The enlightened eye.* New York: Macmillan.

Holliday, W. G. (2002). Selecting a science textbook. *Science Scope*, January, 16-18.

Kesidou, S. (2001). Aligning curriculum materials with national science standards: The role of Project 2061's curriculum materials analysis procedure in professional development. *Journal of Research in Science Teaching*, 38: 47-65.

Kirk, M. Matthews, C. C., & Kurtts, S. (2001). The trouble with textbooks. *Science Teacher*: December, 42-45.

National Research Council. (1999). *Selecting instructional materials.* Washington, DC: National Academy Press.

National Science Resource Center. (1996). *Evaluation criteria for elementary and middle school science curriculum materials.* Washington, DC: National Academy Press.

Stern, L. & Roseman, J. E. (2001). Textbook alignment. *The Science Teacher*, October, 52-56.

CURRICULUM MAPPING

Carr, J. F. & Harris, D. E. (2001). *Succeeding with standards.* Alexandria, VA: Association for Supervision and Curriculum Development.

DeClark, T. (2002). Curriculum mapping: A how-to guide. *The Science Teacher*, April, 29-31.

Jacobs, H. H. (1997). *Mapping the big picture: Integrating curriculum and assessment.* Alexandria, VA: Association for Supervision and Curriculum Development.

Love, N. (2001). *Using data/Getting result: Practical guide for school improvements in mathematics and science.* Norwood, MA: Christopher-Gordon.

Tibbals, C. (2000). *SEIP standards benchmarking.* Washington, DC: Council of Chief State School Officers.

PROFESSIONAL DEVELOPMENT

Allen, D. (1995). *The tuning protocol: A process for reflection.* Providence, RI: Coalition of Essential Standards.

Aronson, E. (1978). *The jigsaw classroom.* Beverly Hills, CA: Sage.

Bennis, E. (1994). *On becoming a leader.* Reading, MA: Addison-Wesley.

de Bono, E. (1982). *De Bono's thinking course.* London: British Broadcasting Corporation.

Garmston, R. & Wellman, B. (2000). *The adaptive school: Developing and facilitating collaborative groups.* Norwood, MA: Christopher-Gordon.

Hopkins, D. (1992). *A teacher's guide to action research.* Philadelphia, PA: Open University Press.

Johnson, A. (2002). *A short guide to action research.* Needham Heights, MA: Allyn and Bacon.

Leedy, P. D. & Ormond, J. E. (2001). *Practical research.* Upper Saddle River, NJ: Prentice-Hall.

Lewin, K. (1948). *Resolving Social Conflicts.* New York: Harper & Row.

Loucks-Horsley, S., Hewson, P. W., Love, N. & Stiles, K. E. (1998). *Designing professional development for teachers of science and mathematics.* Thousand Oaks, CA: Corwin.

Murphy, C. U. & Lick, D. W. (2001). *Whole-faculty study groups.* Thousand Oaks, CA: Corwin.

National Staff Development Council. (2000a). *Revisioning professional development.* Oxford, OH: Author.

National Staff Development Council. (2000b). *Standards for staff development.* Oxford, OH: Author.

National Staff Development Council. (2001). *Tools for growing the NSDC Standards.* Oxford, OH: Author.

North Central Mathematics and Science Consortium. (2000). *Blueprints: A practical toolkit for designing and facilitating professional development* [CD-ROM]. Author.

Ogle, D. (1986). KWL: A teaching model that develops active reading of expository texts. *Reading Teacher, 39,* 564-570.

Parks, S. & Black, H. (1992). *Organizing thinking.* Pacific Grove, CA: Critical Thinking Press.

Pascale, R. (1990). *Managing on the edge.* New York: Simon & Schuster.

Silberman, M. (1996). *Active Learning: 101 Strategies to teach any subject.* Needham Heights, MA: Allyn & Bacon.

Thompson, C. L. & Zeuli, J. S. (1997). The frame and tapestry: Standards-based reform and professional development. In: *The heart of the matter: Teaching as a learning profession.* G. Sykes (Ed.). San Francisco: Jossey-Bass.

PROFESSIONAL DEVELOPMENT DESIGNS

RESOURCE

This Guide is designed primarily for groups of professionals who want to acquire the necessary background for implementing standards-based curriculum and instruction in their classrooms. We have found that the Guide works best when used in conjunction with structured professional development experiences.

Professional Development Designs provide helpful suggestions, time frames, large group focusing questions, and ideas for leaders of groups that are exploring issues associated with standards-based reform. The Guide's approach, and the one that we highly recommend to facilitators, employs constructivist practices that actively and thoughtfully engage educators in investigating issues and applications of standards. We expect that as professional development leaders adapt these materials to their own special audiences, more creative and tailored alternatives will emerge.

> *"Teachers react against ideas and materials that are theoretically sound but do not function in the classroom. They seek proof that other professionals with whom they identify are making new methods work...the fact that others can do it gives them moral support and challenges them."*
>
> **Black & Atkin, 1996**

> *"A successful individual typically sets his next goal somewhat, but not too much above his last achievement. In this way he steadily raises his level of aspiration."*
>
> **Lewin, 1948**

In developing the Guide, we incorporated many principles and ideas recommended for people who lead adult learning groups. In particular, we referenced the National Staff Development Council's (2000b) *Standards for Staff Development* (available at **www.nsdc.org**), *Designing Professional Development for Teachers of Science and Mathematics* (Loucks-Horsley et al., 1998), and the features that Darling-Hammond and McLaughlin (1995) identified for effective professional development. As you adapt and modify the approaches suggested here, you should consider these resources. They provide excellent guidance for program providers.

This Guide employs many pedagogical practices that are potentially transferable to the classroom. We have found that after using these, it is important to ask your audiences, "What are some ways that you could apply this strategy in *your* own practice?" In this way, educators can find direct and ready applications of this professional development experience.

PROFESSIONAL DEVELOPMENT DESIGNS

RESOURCE A

The Guide presumes that users have access to the relevant national and state or local standards documents. We suggest that you prepare a set of transparencies that covers the entire guide, mount it in quick-release covers, and store in a three-ring binder. Some of the templates provided in the Guide may not be usable in their existing form. You may find that these templates and the numerous workspaces for teacher responses need to be adjusted.

We designed this book to be used in its entirety. However, we recognize that few groups of teachers have identical professional development needs and/or matching resources. Through experience, we have discovered that the chapters and activities can be reorganized using a modular approach. Depending on the needs of the audience and program constraints, different materials can be clustered into a variety of configurations. For instance, if there were a need to investigate just the national standards, then the materials in Chapter 2 would be appropriate as a stand-alone resource. Chapter 4 works well to satisfy a request for a single-day session on assessment. In Chapters 6 and 7, we use the concept of a Learning Path to describe the modular method of planning and provide specific examples of this approach. The final section of Professional Development Designs describes how we have applied the Guide's materials for different professional development purposes.

PREFACE TO THE GUIDE

ACTIVITY	WORKSPACE	STEP-BY-STEP	NOTES TO PRESENTER
45 minutes		• Use a transparency to review the Guide's four premises with the entire group. This establishes a clear view of the Guide's central principles and the author's perspectives on standards. • Review Expectations for Users of the Guide on the last page of the Preface. Relate this review to the similar importance of providing clear expectations for students. • Introduce the Learning Cycle (Figure 5.3). Because this instructional planning model provides the organizational framework for the Guide and for the curriculum materials that teachers will develop in Chapter 5, it is important to introduce the model early in the session. • Use the concept map (Figure P.1) to provide an overview of the Guide. Trace the path that your professional development group will be following.	• Group Norms: Consider allocating time to establish group standards of behavior. Include such things as punctuality, listening, limiting "autobiographical" story-telling, openness, etc. Post these on chart paper and display them prominently whenever you meet. • Parking Lot: Have large sticky notes available for teachers to write ideas, concerns and questions that are genuine but not appropriate for this particular moment. Post these on a piece of chart paper and address these issues at the proper time. • This type of PD experience may be new for many teachers. It is normal for them to be anxious and to feel overwhelmed by the expectations and overview.

ENGAGEMENT: THINKING ABOUT STANDARDS AND INSTRUCTION

CHAPTER 1

LEARNING GOALS

- Develop motivated users of the Guide.
- Activate prior beliefs about change.
- Examine personal beliefs and attitudes about standards.
- Explore ideas about constructivist teaching.

ACTIVITY	WORKSPACE	STEP-BY-STEP	NOTES TO PRESENTER
1.1 My Beliefs and Attitudes About Change 45-60 minutes		• Mention that the purpose of an Engagement activity in a learning cycle is to generate interest in the topic. • Give an overview of the activities in this chapter. • Review the directions. Be sure to check for understanding before beginning. Ask one individual in the group to rephrase the task in his or her own words. • Group teachers into pairs. • Use a Think-Pair-Share to process this activity. After completing the reflection as individuals, have participants discuss their responses in pairs. • Use the reflection questions for a large group discussion. Help participants to uncover the central issues that their group must ultimately address during this professional development experience.	This is an ideal time to forge new relationships and to begin building the group into a community of inquirers. • Seasonal Partners (Garmston & Wellman, 2000) is an example of a quick, easy, and active grouping strategy. • Think-Pair-Share: This simple yet highly effective strategy is especially helpful for processing new information. The method facilitates thoughtful discussion about any topic. Directions for this procedure can be found in most reviews of cooperative learning. • Question 4 can be used to introduce the idea of transformative professional development. We recommend that you read the *Frame and Tapestry* paper (Thompson & Zeuli, 1997). This is the type of professional development that the Guide is intended to support.
1.2 Beliefs and Attitudes About Constructivism and Standards-Based Teaching 45-60 minutes		• Review the directions for the survey and the example. Check for understanding before beginning. Ask someone in the group to paraphrase the instructions. • Have people complete the survey and the reflection as individuals. Emphasize that this is a self-assessment and is not meant to evaluate their current beliefs. • Inform teachers that they should be unconcerned if some terms are unfamiliar. All of these ideas are addressed in future chapters of the Guide. • In the large group setting, ask several participants to identify the item that generated the greatest disparity between present and desired conditions. Ask them to explain what they believe accounts for this difference and why they would like to experience change in this area. • Reflection 1.2 reemphasizes the idea that although a changeover to standards-based approaches is neither simple nor direct, it can be accomplished given an appropriate level of commitment.	• Reflection: This is a core strategy for constructivist teaching and learning. Thoughtful reflection is essential for developing a deep and lasting understanding of standards-based reform. Reflections are included with most activities in the Guide. A time-saving approach is to review only the key questions with the entire group. • Matched Pairs: The survey applies this technique. This process illustrates that most human behaviors and knowledge, such as beliefs and attitudes, exist along continua rather than as all-or-nothing phenomena. This approach suggests that change is incremental and thus less daunting. • References and Further Readings: Point these out. Such resources can supplement coverage of topics in the Guide. • Save the results of this survey. Use this inventory as a pre-test/post-test to gauge changes in understanding after completing the Guide. See Evaluation 8.1. • Garmston and Wellman (2000) offer valuable insights about the difference between dialogue and discussion. This is an ideal time to review this distinction because there are numerous opportunities in the Guide for ideas to be considered from a group perspective.

ENGAGEMENT: THINKING ABOUT STANDARDS AND INSTRUCTION

CHAPTER 1

LEARNING GOALS

- Develop motivated users of the Guide.
- Activate prior beliefs about change.
- Examine personal beliefs and attitudes about standards.
- Explore ideas about constructivist teaching.

ACTIVITY	WORKSPACE	STEP-BY-STEP	NOTES TO PRESENTER
SUMMARY REFLECTION: *Thinking About an Issue from the Right-Angle Perspective* 15 minutes		• Assign this Summary Reflection as out-of-class work. Begin the following session by reviewing this reflection. • Review Question 4. Most people have strong views about standards. Mention that it is important for people to be clear about the difference between what they think and what they know.	• Right Angle: This highly visual graphic illustrates the sharp distinction between beliefs and knowledge using a 90-degree angle diagram. • Teaching Tip: Ask participants how they could use the diagram in their own classrooms. What topics would they substitute for Standards and Benchmarks?

EXPLORATION: DISCOVERING THE NATIONAL STANDARDS RESOURCES

CHAPTER 2

LEARNING GOALS

- Assess prior knowledge about standards.
- Develop a language for standards

ACTIVITY		WORKSPACE	STEP-BY-STEP	NOTES TO PRESENTER
2.1	Developing a Common Language About Standards 45 minutes without video 75 minutes if video is shown		• Emphasize that activating prior knowledge is the purpose of the Exploration in a learning cycle. "Getting it right" is not the goal. • Review the activities in this chapter. • Pair teachers with a different Seasonal Partner. See Engagement 1.1. • Be sure to point out the rating scale. Check for understanding. • Ask participants to complete Parts 1 and 2 *before* searching for the matching descriptor. • Assure participants that they will eventually master this language as they complete the Guide. <u>Focusing Questions</u>: • What is the relationship between effective communication and the existence of a common language about a topic? • Ask participants to raise their hands if they learned anything by completing this activity. Ask, "Would it have been more efficient to simply define these terms for you?" Review the distinction between efficient and effective instructional practices. Connect this with the principles of constructivist teaching and learning raised in Engagement 1.2.	• You are likely to find that there is considerable confusion, overlap, and disagreement over these terms. • A good alternative to completing this task is to use a Card Sort. This approach enlivens a particularly drab, but nevertheless important, assignment. • <u>Card Sort</u> (Silberman, 1996): A Card Sort illustrates the practice of getting students actively involved in their own learning. Computer labeling programs make it easy to prepare sets of paired index cards. Place the terms on white cards; use colored cards for the corresponding descriptors. • <u>Teaching Tip</u>: Ask participants if they could use a Card Sort in their own classrooms. For what topics could this approach be useful? • ASCD Video: *Raising Achievement Through Standards*. This contains an overview of standards-based principles. Prepare a set of focusing questions such as a 3-2-1 Reflection, or Sentence Stems (see Summary Reflections 2 and 7 for examples).
2.2	Internet Discovery Guide 15 minutes		• This activity lends itself to an out-of-class assignment. Review it at the next session.	• <u>Discovery Guides</u>: These tools are used throughout the Guide to focus attention on particular concepts. They also streamline the process of gathering data about a topic. • Mention that up to this point in the Guide, standards have only been considered in a generic way. This activity establishes the initial connection with national and state standards.

EXPLORATION: DISCOVERING THE NATIONAL STANDARDS RESOURCES

CHAPTER 2

LEARNING GOALS

- Assess prior knowledge about standards.
- Develop a language for standards.

ACTIVITY	WORKSPACE	STEP-BY-STEP	NOTES TO PRESENTER
2.3 Getting to Know the National Standards 120 minutes		• Use a Jigsaw for this activity. This activity incorporates text-based dialogue. • Create home groups of three. Divide the total number of participants by three. This gives you the number of people in each of the three expert groups. • Use the Discovery Guides to distribute the workload and streamline the review of the national standards documents. Mention that the Discovery Guides include a few thought questions that cannot be directly answered by reading the documents • Each person becomes an expert on one of the national standards resources by working in his or her expert group. Later, people share their understanding with other members of the home group. Use the following sequence for the experts to report out in home groups to: science, pgs. 26-28; mathematics, pgs. 29-31. • Conduct a wrap-up discussion to address lingering issues or questions. • As with all cooperative learning activities, it is important to have participants reflect on the jigsaw process itself. Focusing Question: • What is the principal thing that you discovered from your investigation of the national standards documents?	• This is a very important activity. It gives teachers the opportunity to gain an in-depth familiarity with the resources that they will reference throughout later chapters of the Guide. • Jigsaw (Aronson, 1978): Procedures for this grouping strategy are readily available on the Internet and in books on cooperative learning. • Text-Based Dialogue: The active learning strategy used in Exploration 2.3 focuses discussion around a piece of text. Requiring that specific passages in the material be referenced sharpens attention and facilitates conversation about a topic. • You are likely to find that many teachers are unfamiliar with the national standards documents. • Every national standard resource for mathematics and science has its own Discovery Guide. Create additional Discovery Guides if you want participants to examine other documents (e.g., *Atlas of Science Literacy*). • Assure teachers that in Chapter 3, the Guide's focus shifts to the actual standards that they are responsible for implementing.
SUMMARY REFLECTION: *Realizing the Promise of a Standards-Based Education*		• Assign this as an out-of-class reading. • Review the reflection at the next session.	• Sentence Stems: This is a focused reading strategy. Stems are prompts that direct the analysis of text. • Teaching Tip: Ask participants how they could use Sentence Stems with their students.

EXPLANATION I: RESEARCHING NATIONAL RESOURCES TO CLARIFY LEARNING GOALS

CHAPTER 3

LEARNING GOALS

- Examine how a Content Clarification can foster an in-depth understanding of a standard or learning goal.
- Construct a Content Crosswalk.
- Recognize how national standards resources can provide information for interpreting state or local content standards.

ACTIVITY	WORKSPACE	STEP-BY-STEP	NOTES TO PRESENTER
3.1 Content Clarification of a National Standard 120 minutes		• Mention that the purpose of an Explanation in a learning cycle is to introduce content information. • Review the activities in this chapter. • Have teachers work in the same groups of three that were used for Exploration 2.3. Each team member will complete one of the three columns in the Content Clarification. • Distribute extra copies of Figure 3.1. • Allow ample time to study the Content Clarification procedure. Check for understanding before beginning. Structure sufficient time for participants to share their findings. • Prior to reviewing Reflection 3.1, use the example in Figure 3.2 to illustrate a complete Content Clarification. • Mention that this process is essentially a "preplanning" step that clarifies the ends (standards) toward which all curriculum and instruction should ultimately be directed. Focusing Question: • What is the greatest potential value of a Content Clarification?	• *This is the single most important activity in the Guide. Give teachers however much time is needed to complete it successfully.* • Have one or two participants explain the Content Clarification procedure in their own words. This should give you a clear indication of the level of understanding within the group. • This procedure demonstrates that national standards documents have important and practical applications for teachers. • The scaffolded set of questions helps teachers to identify a relevant standard to study. Use both national documents for science and mathematics. • Caution: Expect participants to express reservations, concerns, confusion, and frustration with the Content Clarification. Most teachers eventually grasp the significance of the process. Be sure to carefully monitor progress. • Consider copying and distributing the resulting small group Content Clarifications. This illustrates the relative ease with which a compendium of Content Clarifications can be assembled.
3.2 Creating a Content Crosswalk 75-90 minutes		• Part 1: In the large group, walk teachers through the scaffolded questions for analyzing Figure 3.3. Check for understanding. • This is a good time for organizing participants into the work groups for completing Chapters 4-5. • Be sure to define what a "sighting" means. You can demonstrate the method for identifying a sighting by using one of the boxes in Figure 3.3 as an example. • If participants have difficulty selecting a learning goal for their unit, have them repeat the procedure found in Exploration 2.4. • Distribute extra copies of Figure 3.4.	• This activity increases awareness of the K-12 learning goals associated with a particular standard. Teachers may question the need for considering K-12 standards. Explain that this perspective illustrates where students have been and where they are headed. It also raises the point that helping students to meet standards is a responsibility shared by *all* educators. • The product of the Crosswalk is a framework that is necessary for completing the Content Clarification procedure in the next activity.

EXPLANATION I: RESEARCHING NATIONAL RESOURCES TO CLARIFY LEARNING GOALS

CHAPTER 3

LEARNING GOALS

- Examine how a Content Clarification can foster an in-depth understanding of a standard or learning goal.
- Construct a Content Crosswalk.
- Recognize how national standards resources can provide information for interpreting state or local content standards.

ACTIVITY	WORKSPACE	STEP-BY-STEP	NOTES TO PRESENTER
3.2 Creating a Content Crosswalk (continued) 75-90 minutes		• Point out the new column in Figure 3.4 for entering sightings located in a teacher's "Principal Classroom Resource." This will begin to connect the Crosswalk directly to the teacher's own classroom. <u>Focusing Question</u>: • What is the greatest potential value of a Content Crosswalk?	• Expect participants to express doubts and concerns with connecting the Crosswalk procedure to their own practice. They may not appreciate the importance of gaining a thorough and in-depth understanding of standards and learning goals as a prerequisite for designing instruction or evaluating curriculum. • It is important to fully process this Crosswalk experience with participants. This is the first tangible link between the Guide and their own teaching.
3.3 Applying a Content Crosswalk to Clarify a Learning Goal 45 minutes		• Emphasize that this Content Clarification will only address a single learning goal from the previous activity. • Ask teachers to identify the new column that appears in Figure 3.5. Ask if they think it is an appropriate time to begin thinking about assessment. • Distribute extra copies of Figure 3.5. <u>Focusing Question</u>: • Could you combine the Content Crosswalk and Content Clarification into a single process?	• The product of this Content Clarification is very important. Participants use this information as the basis for their performance assessment in Chapter 4 and the accompanying instructional unit in Chapter 5. • <u>Caution</u>: Specific information about assessing student work is not available from the national resources listed in the Crosswalk template. Teachers may include insights derived from their own "craft knowledge" (i.e., personal experiences) in both assessment and teaching columns. Several sources of information about assessment are cited in References and Further Readings.
SUMMARY REFLECTION: *Content Clarification of Standards and Learning Goals* 15 minutes		• Assign this as out-of class work. Review at the next session. • A nice out-of-class assignment is to have teachers reread the quotes that appear in the Guide. Ask them to select the quote that they find most meaningful and to explain why they selected this particular item. • You can use the quote task anytime. For discussion purposes, cut and paste to make a collection of these quotes on transparencies.	

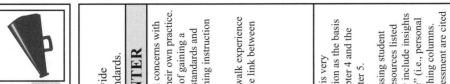

EXPLANATION II: ASSESSMENT IN A STANDARDS-BASED SYSTEM

CHAPTER 4

LEARNING GOALS

- Investigate connections among learning goals for students, assessment, and the instructional practices of teachers.
- Examine prior understanding of standards-based assessment.
- Develop a high level of familiarity with terminology associated with assessment.
- Recognize how national standards resources can provide information for interpreting state or local content standards.
- Design a performance assessment that targets a particular learning goal.

ACTIVITY	WORKSPACE	STEP-BY-STEP	NOTES TO PRESENTER
4.1 Congruence With Standards 60 minutes		• Review the activities in this chapter and remind teachers that the purpose of an Explanation is to increase content knowledge. • Assign the Guided Reflection in Chapter 4 as a Paired Reading for recapping the first three chapters. • Lead a large group discussion to summarize the major discoveries that teachers have made since using the Guide. • Have teachers work in groups of four to complete the Congruence activity. Retain these groups for the next activity. • Take time to complete a thoughtful discussion of Figure 4.3. Focusing Question: • Why is congruence such an important concept when discussing standards-based reform issues?	• Paired Reading (Garmston & Wellman, 2000): This is a focusing strategy. Designate team members as A and B. Have team members alternate in reading successive paragraphs aloud. Have listeners paraphrase the passages. Continue in this manner until the Guided Reflection is completed. • Julian's experience as described in the Guided Reflection should mirror that of the typical user of the Guide. • Understanding the Congruence Triangle is critically important for developing a rationale for standards-based teaching and learning.
4.2 Traditional Versus Standards-Based Instruction 30 minutes		• In the large group setting, have teachers identify the distinctive features of standards-based practice. • Briefly allude to Backward Design. Emphasize that it is not the components of instructional planning that differ in backward design, but rather their placement and the amount of attention makes this process distinctive. Focusing Question: • What is the greatest challenge that you expect to experience as you implement standards-based curriculum and instruction?	• By this point, participants are likely to be eager to consider topics other than standards! • Although this diagram depicts traditional and standards-based practices as separate and distinct, this is an artificial dichotomy, because the two approaches generally overlap. • Backward Design: (Wiggins & McTighe, 1998). According to this method, instructional planning progresses from standards to assessment to the development of curriculum. This approach provides the conceptual framework for developing a standards-based unit in Chapter 5. • Venn Diagram: This strategy focuses attention on the essential attributes of a subject using a compare-and-contrast approach.

EXPLANATION II: ASSESSMENT IN A STANDARDS-BASED SYSTEM

CHAPTER 4

LEARNING GOALS

- Investigate connections among learning goals for students, assessment, and the instructional practices of teachers.
- Examine prior understanding of standards-based assessment.
- Develop a high level of familiarity with terminology associated with assessment.
- Recognize how national standards resources can provide information for interpreting state or local content standards.
- Design a performance assessment that targets a particular learning goal.

ACTIVITY	WORKSPACE	STEP-BY-STEP	NOTES TO PRESENTER
4.3 The Language of Assessment 45 minutes		• From this point forward, it is best for teachers to be organized into work pairs. • You have two choices here: process the activity as presented in the book or use a Card Sort (see Exploration 2.1). • If you use a Card Sort, provide these instructions: • Sort 1: White cards contain terms that are commonly associated with assessment. Colored cards contain descriptions of these assessment terms. Try to match the terms with the matching textbook definition. Do not be concerned if you aren't able to find some matches. • Sort 2: Separate the matched pairs from Sort 1 into two stacks. In the first pile, place those aspects of assessment that directly impact your life as a professional. Place those assessment items that have little or no impact in the second pile. Focusing Question: • Why has the subject of assessment risen to such a high level of importance in educational circles?	• We designed active learning strategies for exploring the topic of assessment. However, this section presents a large amount of content information. If this material is new for a teacher, it tends to be confusing. Monitor progress carefully and provide direct assistance as needed. • This activity establishes a common language for the terminology used to discuss assessment. • Question 4 in the reflection for Explanation 4.3 requires some independent research on the Internet. The distinction between formative and summative assessment is one that teachers need to understand. • Questions that arise about standardized testing should be temporarily placed in the Parking Lot. These can later be addressed in conjunction with the second reading at the end of this chapter. • One advantage for using the procedure as given in the Guide is that it requires more thought. Teachers will often discover that they know more about assessment than they formerly believed.
4.4 Elements of a Performance Assessment 30 minutes		• Ask teachers to carefully study Figure 4.7. Use a Think-Pair-Share to process Part A. • Walk teachers through the Discovery Guide for analyzing the performance assessment. • Use transparencies and frequently check for understanding during this activity. Focusing Question: • Why is this type of assessment sometimes referred to as "authentic"?	• Explanation 4.4 comes as close to direct instruction as any section of the Guide. • This activity provides background information about performance assessments and introduces the idea of using a performance (product or action) to determine if students are meeting a standard. • Performance assessment is the central feature of a standards-based assessment system. Be sure that the group has a firm grasp of the relevant terminology.

EXPLANATION II: ASSESSMENT IN A STANDARDS-BASED SYSTEM

CHAPTER 4

LEARNING GOALS

- Investigate connections among learning goals for students, assessment, and the instructional practices of teachers.
- Examine prior understanding of standards-based assessment.
- Develop a high level of familiarity with terminology associated with assessment.
- Recognize how national standards resources can provide information for interpreting state or local content standards.
- Design a performance assessment that targets a particular learning goal.

ACTIVITY	WORKSPACE	STEP-BY-STEP	NOTES TO PRESENTER
4.5 Building Your Performance Assessment 60 minutes		• Begin by posing the question found in the introduction to this activity. • Continue in the small work groups for Part A. Ask teachers to carefully study Figure 4.9. Emphasize that Design Details for a Performance Task is an elaboration of the first two parts of the performance assessment described in Explanation 4.4. The third piece, a scoring rubric, is developed in Explanation 4.6. • Use a Think-Pair-Share to process Part A (see Engagement 1.1). • For Part B, refer teachers to their Content Clarification of a Learning Goal from Explanation 3.3. • Review the sample performance task in Figure 4.10. Be sure that teachers understand the six elements listed in Part B. • Provide the opportunity for teachers to share their draft performance assessments and to have them critiqued by colleagues using Warm and Cool feedback. Focusing Question: • What was the major difficulty you encountered when designing the performance task?	• This activity uses the findings from the Content Clarification as the basis for developing a performance assessment that is congruent with the selected learning goal. • Processing the sample performance task in Figure 4.10 as a large group is very important. Have one or two participants explain the figure in their own words. This should give you a clear indication of the level of understanding within the group. • This task can be confusing. Teachers generally grasp the principles of performance assessment during Explanation 4.4. However, they often experience difficulty inventing an appropriate task and identifying the corresponding criteria for success. • Warm and Cool Feedback: This procedure is adapted from the Tuning Protocol developed by the Coalition of Essential Schools. The process enables a person to receive feedback from peers in a non-threatening, constructive manner. • In this case, the Tuning Protocol feedback should reveal whether teachers have successfully applied the six essential elements of the performance task.
4.6 Rubrics for Scoring Student Work 90-120 minutes		• Mention that most rubrics fall into one of the two categories identified in Part A. Have teachers use the Thinking Diagram to distinguish between the different types. • Working with the large group in Part B makes it easier to develop a common language for rubrics. Before continuing, check for understanding by using the questions in this section.	• Compare and Contrast Thinking Diagram: This graphic organizer is an analytical tool that matches items according to their distinctive attributes. • Teaching Tip: Ask participants how they could use a Compare and Contrast Diagram with their students. • A common misconception is to classify a rubric as an assessment. A rubric is the tool through which the assessment is scored.

EXPLANATION II: ASSESSMENT IN A STANDARDS-BASED SYSTEM

CHAPTER 4

LEARNING GOALS

- Investigate connections among learning goals for students, assessment, and the instructional practices of teachers.
- Examine prior understanding of standards-based assessment.
- Develop a high level of familiarity with terminology associated with assessment.
- Recognize how national standards resources can provide information for interpreting state or local content standards.
- Design a performance assessment that targets a particular learning goal.

ACTIVITY	WORKSPACE	STEP-BY-STEP	NOTES TO PRESENTER
4.6 Rubrics for Scoring Student Work (continued) 90-120 minutes		• Part C can be completed on the Internet or with blank rubric templates that you provide. Instruct teachers to carefully connect the rubric with the Criteria for Success developed in Explanation 4.5. <u>Focusing Question</u>: • What lingering issues and concerns do you have about performance assessment?	• If your group lacks Internet access, printed rubric templates must be provided. • By using the Scoring Guide for Evaluating a Rubric (Figure 4.14), participants gain additional experience with these scoring tools. • Rubric development software is relatively inexpensive and simplifies the process of creating and saving rubrics.
SUMMARY READINGS AND REFLECTIONS: *1. Facing and Embracing the Assessment Challenge* *2. The Authentic Standards Movement and Its Evil Twin* 30 minutes		• Use these readings as out-of-class assignments.	• The second Reflection highlights the confusion between standards-based assessment and standardized testing. This issue is sure to arise and should be addressed even if this reading is not assigned.
GUIDED REFLECTION 15 minutes		• Use a transparency for the Guided Reflection at the end of Chapter 4 to review the available Extension options. Be sure that teachers clearly understand the three different alternatives. They should make their choice before proceeding. <u>Focusing Question</u>: • What Extension have you chosen to follow and why?	• <u>Caution</u>: It is important for the group to resolve which of the Extensions it wishes to pursue. Participants may want to divide into different subgroups. Chapter 5 is the Extension option that most groups will select.

EXTENSION I: DEVELOPING STANDARDS-BASED CURRICULUM MATERIALS

CHAPTER 5

LEARNING GOALS

- Identify the essential features of standards-based instruction.
- Finalize the performance assessment.
- Make necessary lesson accommodations for all learners.
- Select learning experiences that target a particular learning goal.

ACTIVITY		WORKSPACE	STEP-BY-STEP	NOTES TO PRESENTER
5.1	Refining Your Performance Assessment Varies		• Review the role of an Extension in a learning cycle. Extensions provide opportunities to apply new knowledge and skills. • Review the activities in this chapter. Focusing Question: • What was the connection between this survey instrument and the principles of effective performance assessment found in Chapter 4?	• No time estimates are provided for the activities in this chapter. Some groups may only wish to review the templates; others may want to complete the standards-based unit within the sessions. We recommend that teachers actually develop a full unit based on the standards-based principles of the Guide. • This self-study tool provides a systematic way to examine a performance assessment for consistency with the principles introduced in Chapter 4.
5.2	Ensuring That Learning Experiences Are Accessible to All Students Varies		• Point out to teachers that this step is ongoing and should be incorporated into all phases of their instructional planning. Focusing Question: • What is the connection between this activity and the statement, "Standards are for all students?"	• The Student Activities Profile Model is a reflective tool that focuses attention on the accommodations necessary for making instructional units equally accessible to all students.
5.3	Designing Instruction: Standards-Based Templates Varies		• Have teachers work in their small groups to complete their instructional unit. • Review the learning cycle diagram and instructional unit templates before beginning. It is important to check for understanding before teachers start developing their unit. • Ask the group to establish a realistic timetable for completing this project. • Point out to teachers that the templates may have to be restructured to accommodate their materials. • Mention that the cover sheet is essentially an *abstract* for their unit. • The units that teachers develop should be presented to the entire group. Use a Tuning Protocol to guide the presentations.	• Have one or two participants explain the unit design procedure in their own words. This should give you a clear indication of the level of understanding within the group. • Because teachers are well grounded in the standards and have already developed their performance assessment, this section can move quickly. • As they develop their units, teachers may find themselves reverting to the familiar traditional model described in Figure 4.4. Emphasize that it is *essential* to follow the standards-based model. • Your role at this point is to provide helpful assistance.

EXTENSION I: DEVELOPING STANDARDS-BASED CURRICULUM MATERIALS

CHAPTER 5

LEARNING GOALS

- Identify the essential features of standards-based instruction.
- Finalize the performance assessment.
- Make necessary lesson accommodations for all learners.
- Select learning experiences that target a particular learning goal.

ACTIVITY	WORKSPACE	STEP-BY-STEP	NOTES TO PRESENTER
5.3 Designing Instruction: Standards-Based Templates (continued) Varies		Focusing Question: • What do you consider to be the most rewarding aspect of standards-based instructional design?	• Tuning Protocol (Allen, 1995): This is a structured way for teachers to share their standards-based units with colleagues. Public displays of work with guided feedback are powerful forces for developing collegiality within professional development teams. Although this process is time-consuming, it ensures that adequate attention is directed to the work product of teachers. Instructions for the protocol are readily accessible on the Internet. • Teaching Tip: Ask participants how they could use a Tuning Protocol with their students.
SUMMARY REFLECTION: *Self-Assessment of Standards-Based Materials* Varies		• Individual work teams should complete this self-assessment. • If possible, review this reflection with the entire group after all of the work teams have presented their units.	

EXTENSION II: ANALYZING AND EVALUATING CURRICULUM MATERIALS

CHAPTER 6

LEARNING GOALS

- Recognize how ideas about national standards, constructivist teaching and learning, performance assessment, and curriculum design provide a framework for analyzing instructional materials.
- Apply the Guide's major principles to analyze curriculum materials.
- Utilize the Curriculum Analysis Rubric.
- Interpret data from the Curriculum Analysis Rubric to evaluate curriculum materials.

ACTIVITY	WORKSPACE	STEP-BY-STEP	NOTES TO PRESENTER
6.1 Preparing to Analyze Curriculum Materials Varies		• Review the role of an Extension in a learning cycle. Extensions provide opportunity to apply new knowledge and skills. • Review the activities in this chapter. • Ask the group to establish a realistic timetable for completing this project. • Use the Guided Reflection to create a context for your curriculum analysis project. <u>Focusing Question</u>: • Why is adequate preparation a necessary precursor for effectively analyzing curriculum materials?	• No time estimates are provided for the activities in this chapter. Completing this process according to the prescribed procedure requires thoughtful deliberation and a considerable amount of time. • If you are using the Guide strictly to direct a curriculum analysis initiative, then you should have teachers complete the seven-step Learning Path before beginning. • Following the prescribed Learning Path is a necessary prerequisite because the curriculum analysis procedure is anchored in the ideas introduced during these activities. • Curriculum analysis and evaluation can be a powerful professional development experience.
6.2 Curriculum Analysis and Evaluation Varies		• Mention that this is a standards-based analysis and evaluation of curriculum materials. • In the large group, carefully review the directions for Part A, the sighting sheet, and the scoring rubric. • Proceed only when you are confident that the group has reached a full and consistent understanding of the process. • Create work teams of two for looking at individual curriculum materials. • Data drawn from Part A provides the basis for evaluation. Parts B-D synthesize the findings and conclude with final recommendations. <u>Focusing Question</u>: • What is the chief advantage of completing a systematic and standards-based analysis and evaluation of curriculum materials?	• Have one or two participants explain the curriculum evaluation process in their own words. This should give you a clear indication of the level of understanding within the group. • Some groups do not feel comfortable with the idea of using one or two learning goals as the basis for making decisions about curriculum materials. This approach, however, is helpful for dividing the labor within larger groups. • You should adjust this process so that the group as a whole is satisfied that it will accomplish the desired goals.
SUMMARY REFLECTION: *Standards-Based Curriculum Evaluation*		• Individual work teams should complete this reflection. • If possible, review this reflection after the entire curriculum analysis and evaluation group presents its findings.	

EXTENSION III: CURRICULUM MAPPING

CHAPTER 7

LEARNING GOALS

- Apply a systematic procedure for mapping a curriculum.
- Develop a K-12 curriculum framework that is aligned with the relevant standards.

ACTIVITY	WORKSPACE	STEP-BY-STEP	NOTES TO PRESENTER
7.1 Preparing to Map the Curriculum Varies		• Review the role of an Extension in a learning cycle. Extensions provide opportunity to apply new knowledge and skills. • Review the activities in this chapter. • Ask the group to establish a realistic timetable for completing this project. Focusing Question: • Why is adequate preparation a necessary precursor for effectively mapping a K-12 curriculum?	• If you are using the Guide strictly to direct a curriculum mapping initiative, then you should have teachers complete the five-step Learning Path before beginning. • Following the recommended Learning Path is a prerequisite because the mapping procedure is anchored in the ideas introduced during these activities. • No time estimates are provided for the activities in this chapter. Goals established by the mapping committee will determine the depth to which your group probes into its curriculum. • The procedure works best with groups of K-12 teachers that are organized by content area and/or grade level. • Any educational group that is committed to standards-based reform will, of necessity, complete some degree of inter- and intra-grade-level analysis.
7.2 Curriculum Mapping Varies		• In the large group, carefully review the directions for Part A. Be sure that teachers understand how to use the mapping template. • Review the match-gap analysis procedure. • For Part B, create work teams consistent with the particular makeup of the group. To ensure consistency, have the team members carefully review the Scoring Variations. Clarify the distinction between addressing a learning goal through instruction and assessing student performance for a learning goal. • Proceed only when you are confident that the group has achieved consensus understanding about the process. • Point out that the mapping templates may have to be adjusted to accommodate the particular needs of the team. Focusing Question: • What is the chief advantage of completing a systematic and standards-based process of curriculum mapping?	• Have one or two participants explain the mapping process in their own words. This should give you a clear indication of the level of understanding within the group. • Mapping is the means for gathering data. Developing and implementing the Action Plan are the methods for applying this data. A curriculum that is aligned with standards for students is the desired end product. • This mapping procedure is rigorous. The analysis has the potential for producing significant changes in the delivery of K-12 curricula. • Local curricula can be mapped against state or national standards. Whatever standards are selected should be incorporated into templates similar to the Tennessee model in Figure 7.1. • The basic framework in Figure 7.8 includes the generic features most commonly recommended for inclusion in Action Plans. The format can be modified according to the special requirements of the mapping team.
SUMMARY REFLECTION: *3-2-1 Curriculum Mapping*		• Individual work teams should complete this reflection. • If possible, review this reflection after the entire curriculum mapping team presents its findings.	• 3-2-1 Reflection: This is another focusing strategy that differentiates responses according to ideas, actions, and insights. • Teaching Tip: Ask participants how they could use a 3-2-1 Reflection with their students.

EVALUATION: DISCOVERIES ABOUT STANDARDS-BASED TEACHING AND LEARNING

CHAPTER 8

LEARNING GOALS

- Analyze changes in beliefs and understanding.
- Assess what was learned through completing the major parts of the Guide.
- Evaluate the impact of the Guide.

ACTIVITY		WORKSPACE	STEP-BY-STEP	NOTES TO PRESENTER
8.1	Understandings About Constructivism and Standards-Based Learning		• Review the role of an Evaluation in a learning cycle. Evaluations provide the opportunity for everyone to assess the amount and type of learning that has occurred during an educational experience. • Review the activities in this chapter. • Have all teachers complete Evaluation 8.1 because it compares initial and final understanding of fundamental constructivist and standards-based reform principles. • All remaining Evaluations are optional. The way that you employ them depends on how you used the various sections of the Guide and what you regard as the best method for wrapping up this professional development experience. The review will also be influenced by which Extension was selected.	• No time estimates are provided for the evaluation activities. • Your principal role is as the facilitator for reviewing the reflection questions. • If you do not have a formal professional development evaluation instrument, you can use Evaluation 8.5 to provide you with feedback. Suggest that teachers send copies of their comments to the authors. • <u>Plus/Minus/Interesting Reflection</u> (de Bono, 1982): Evaluation 8.4 is a focusing strategy for considering an issue from the perspective of its advantages and disadvantages. • <u>Teaching Tip</u>: Ask participants how they could use a PMI reflection with their students.
8.2	National Standards as Resources for Instruction			
8.3	Assessment That Guides Teaching and Learning			
8.4	Developing and Evaluating Standards-Based Materials			
8.5	Standards in the Classroom: An Implementation Guide			
(clock)	Varies			

PROFESSIONAL DEVELOPMENT APPLICATIONS OF THE GUIDE MATERIALS

PURPOSE	AUDIENCE	PROFESSIONAL DEVELOPMENT FORMAT	CHAPTERS FROM THE GUIDE	CONTENT AREAS
Support standards-based reform initiative	K-12 teachers. Warwick (RI) Public Schools.	Three-credit college course. Twelve three-hour sessions.	1-5, 8	Mathematics, Science, English, History, Physical Education, Art
Support standards-based curriculum and instruction	Eisenhower Grant. Rhode Island Board of Governors of Higher Education. K-12 standards institute.	Three-credit college course. Five-day summer institute.	1-5, 8, 9	Science
Support standards-based reform leadership initiative	K-12 teachers and administrators. Bridgewater/Raynham (MA) Public Schools.	Five-day summer institute.	1-5	Mathematics, Science, English, Social Studies, Health Education
Support NSF standards-based reform leadership initiative	Middle school teachers. HELMSS Project. Rhode Island College.	Four-day institute.	1-6, 8	Science
Support standards-based reform leadership initiative	K-16 educators and administrators. Appalachian Regional Education Lab.	Two-day workshop.	1-3	Science
Support standards-based curriculum and instruction	High school teachers. Bristol/Warren (RI) Public Schools.	Six three-hour workshops.	2-5	Science
Disseminate Content Crosswalk and Content Clarification tools	Tennessee teachers. TN Department of Education. Vanderbilt University.	CD-ROM.	3	Science
Prepare for state textbook adoption	Tennessee K-12 Textbook Adoption Committee. Memphis Public Schools.	Five-hour workshop.	6	Science
Prepare for local textbook adoption	Textbook Adoption Committee. Memphis (TN) Public Schools.	Two-day workshop.	6	Science and Mathematics
Overview of national and Tennessee standards	Eisenhower Grant. Tennessee middle school teachers.	Half-day workshop.	1-3	Science
Overview of standards-based reform	Massachusetts Association of Science Supervisors.	One-hour presentation.	2, 3	Science
Overview of national standards resources	Master of Arts in Teaching (Biology) students. Education Department, Brown University	Three-hour presentation.	2, 3	Science
Mapping education program curriculum	College faculty. School of Education, Roger Williams University.	Department meetings, individual work.	7	N/A
Investigate standards-based curriculum and instruction	Secondary preservice teachers. Roger Williams University.	Four-credit semester course.	1-5, 6, 8	Mathematics, Science, English, History, Foreign languages
Provide instruction about standards and assessment	Elementary preservice teachers. Roger Williams University.	Several three-hour sessions.	2, 4	All areas
Provide instruction about standards	Graduate preservice secondary teachers. University of Tennessee.	Two three-hour sessions.	1-3	Science

INDEX

action plan • 123, 131, 150
action research • 147, 149
alignment • 7, 8, 52, 87, 107, 115, 119, 120-123, 131, 142, 154
analytic rubric • 73
assessment • 43, 61, 139-141, 150, 166-169
assessment principles • 58, 60-62, 64, 67, 69, 79, 80, 133, 139, 141, 152, 153, 167-171, 174
Atlas for Science Literacy • 29, 30, 34, 36, 163

backward design • 54, 99, 166
belief inventory • 3, 160
benchmark • vi, 9, 18, 20, 24, 27, 29, 30, 37, 65, 119, 138, 152,
Benchmarks for Science Literacy • 20, 24, 29, 30, 36, 70
best practice • 148
bricolage • 142

change (teacher) • x, 1-4, 133, 145, 151, 160
congruence • 51, 53, 90, 104, 107, 110
congruence triangle • 51, 67, 166
constructivism • xi, 1, 5, 28, 101-104, 134, 136, 150, 157, 160-162, 172, 174
content area • 28, 37
content clarification • 29-33, 45, 48, 50, 107, 118, 138, 164, 165
content crosswalk • 27, 34, 35, 38, 42, 49, 138, 164
content knowledge • 33
correlation • 52, 53
criteria for success • 63, 64, 67
curriculum analysis • 107, 108, 110, 111, 125, 172
curriculum crosswalk • 36
curriculum mapping • 115, 116, 123, 131, 132, 173
curriculum materials • 101, 103, 105, 108, 113, 150, 172

descriptor • 12, 75, 108, 162

engagement • xi, 8, 134, 160
evaluation • xii, 83, 107, 108, 111, 133, 138, 139, 172, 174
explanation • xii, 43, 164
exploration • xii, xiv, 11, 162
extension • xiv, 85, 150, 169, 170, 173

gaps • 117, 119, 130
grade ranges (clusters) • 37, 41, 43, 103, 119, 126, 129

holistic rubric • 72

Individuals with Disabilities Education Act (IDEA) • 88
inquiry • xiii, 6, 21, 135, 146-149, 151
instructional planning • 54, 55, 67, 90, 91

Internet • 11, 16, 17, 60, 71, 79, 81, 149, 162, 163, 169, 171

Journal of Research in Science Teaching • 7

learning cycle • xi, 1, 90, 92, 150, 159, 170
learning goal • 27, 34, 37, 38, 43, 104, 107, 108
learning path • xi, 103-107, 112, 117, 118, 158, 173

matches • 125, 127, 128

National Research Council • 60, 103
National Science Education Standards • 21, 29, 70
National Staff Development Council • 123, 148
North Central Regional Educational Laboratory • 139

performance assessment • 13, 62, 63, 67, 69, 167, 168, 170
performance criteria • 75
performance indicator • 110
performance level • 75
performance task • 13, 63, 64, 67-69, 87, 168
PMI Reflection • 142, 174
Principles and Standards for School Mathematics • 22, 30
professional development • xii, 146, 148, 157, 160, 175
Project 2061 • 30

rubric • 13, 71, 75, 76, 99, 102, 107, 108, 110, 112, 168, 169
rubric generator • 76, 169

scaffolding • 28
Science for All Americans • 19, 23, 29, 30
scoring guide • 64, 66, 69, 76, 87
scoring system • 63
self-assessment • 5, 13, 71, 79, 87
self-study • 170
sightings • 35, 43, 108, 164
standardized testing • 80-82
standards-based curriculum • 7, 133, 147
standards-based instruction • 62, 101, 145, 147
standards movement • ix, x, 81-83, 107, 133
standards resources • 45, 104, 133, 138, 158, 163, 168, 174
Student Activities Profile Model • 88, 89, 170

Tuning Protocol • 168, 170, 171

validity • 52, 53
variations • 75, 76, 110